Lecture Notes in Mathematics

A collection of informal reports and seminars
Edited by A. Dold, Heidelberg and B. Eckmann, Zürich

252

David A. Stone

Massachusetts Institute of Technology, Cambridge, MA/USA
and
State University of New York at Stony Brook, Stony Brook, NY/USA

Stratified Polyhedra

Springer-Verlag
Berlin · Heidelberg · New York 1972

AMS Subject Classifications (1970): Primary: 57 B05, 57 C40, 57 C50
Secondary: 55 F65, 57 F20

ISBN 3-540-05726-9 Springer-Verlag Berlin · Heidelberg · New York
ISBN 0-387-05726-9 Springer-Verlag New York · Heidelberg · Berlin

Offsetdruck: Julius Beltz, Hemsbach/Bergstr.

Introduction

The present paper is a revised version of a much longer work (which had fewer results) entitled "Block Bundle Sheaves". This is an obsolete phrase for "stratified polyhedra"; and since my work has been referred to under the old title, I mention it as a sort of subtitle of the present work. During the process of revision, I profited greatly from lecturing in a graduate seminar at M.I.T., since it was necessary to describe technical definitions and procedures intuitively. I talked about my difficulties to many friends, especially Ralph Reid, my sister Ellen and Dennis Sullivan. Above all, I am indebted to Colin Rourke, who suggested the present method of defining stratifications, and who urged me to change terminology to conform better with Thom's theory [28] of "ensembles stratifiés". I am also glad to express my gratitude to William Browder.

The two facts about block bundles, as defined by Rourke and Sanderson, which make them so important are:

1. if M is a p.ℓ. submanifold of Q, then M has a normal block bundle in Q, which is in some sense unique;

2. there is a relative transversality theorem: if M, N are submanifolds of Q, then there is an arbitrarily small isotopy f_t of Q such that $f_1 N$ is block transverse to M; moreover if bdy N is already block transverse to bdy M in bdy Q, then we may require f_t to be the identity on bdy Q.

That M has a block bundle neighbourhood in Q was shown by a simple construction: roughly speaking, take a simplicial triangulation B of Q in which M is covered by a full subcomplex A. Let B' be a first derived subdivision of B. For each simplex $s \in A$, let

$$s*B = \hat{s} * \ell k(s, B') \text{ (the simplicial join);}$$

$$s*A = \hat{s} * \ell k(s, A')$$

where $\hat{s} \in B'$ is the barycentre of s. Then s*B is a block over s*A, and $\{s*B : s \in A\}$ forms the required block bundle.

This construction can equally well be performed when M and Q are no longer manifolds but general polyhedra. It seemed reasonably to hope that by following the proofs of Rourke and Sanderson one could prove

1. some sort of uniqueness result, and
2. a relative block transversality theorem.

It is the main purpose of the present paper to carry out this program.

In Chapter 1 we give an abstract codification of the kind of structure offered by our construction. There are three stages of the analysis. First we group together all points of M that "look alike" locally in the pair (Q, M). This expresses M as a disjoint union of open sets $\mathcal{M}_i$, each of which is a manifold and in fact a union of open simplexes of A. The family $\{\mathcal{M}_i\}$ is called a "variety" of M in Q. It follows that each closure cl $\mathcal{M}_i$ is a subcomplex A_i of A. Our construction thus provides us with regular neighborhoods of A_i in Q, in M and in each A_j (by Cohen's definition of a regular neighbourhood [7]). These regular neighbourhoods fit together tidily enough, and we describe the situation in terms of a "normal regular neighbourhood system" for the variety $\{\mathcal{M}_i\}$ of M in Q. Next we note that $\mathcal{M}_i$, in case Q is a manifold, is a submanifold of Q, and that our construction puts on (very roughly) the regular neighbourhood of A_i in Q the structure of a normal block bundle -- this is Rourke and Sanderson's construction. This suggests that the regular neighbourhoods of A_i in the general Q, M and the various A_j can also be given block bundle structures, if we allow block bundles whose fibre is no longer a disk but any cone. Our final step is to define such block bundles and describe how they are to fit together. This completes our definition of a "stratification" of M in Q. It should be observed that our construction for M in Q includes a construction for M in itself. Thus our final stratification $\underline{\xi}$ of M in Q includes a stratification $\underline{K}$ of M. It is fruitful to think of $\underline{\xi}$ as a "generalized bundle" over K; this is why I use the notation $\underline{\xi}/\underline{K}$ with Greek letters for the "bundle" space and Roman letters for the "base" space.

In Chapter 2 we simply verify that the result of our construction satisfies the axiomatic systems of Chapter 1.

Theorems about stratifications are proved by dealing inductively with one block bundle at a time. (It is a meta-theorem that any geometrical proof in Rourke and Sanderson's [19, I and II] applies to a block bundle with arbitrary fibre whose base is a <u>manifold</u>. This idea is developed in the Appendix; see also p. vii.) In fact, over each $\mathfrak{M}_i$, its block bundles in Q, M and so on, fit like sub-bundles one of another, and can be treated simultaneously. Difficulties arise when passing from block bundles over some $\mathfrak{M}_i$ to block bundles over some other $\mathfrak{M}_j$. In proving the uniqueness theorems of Chapter 3 for stratifications of fixed M in Q, these difficulties are overcome by a massive use of induction on "complexity" (that is, the number of terms in the variety). I should add that the uniqueness of normal regular neighbourhood systems of M in Q is a straightforward consequence of the usual uniqueness theorem for regular neighbourhoods. We also prove that if $\underline{K}$ is a fixed stratification of the variety $\{\mathfrak{M}_i\}$ of M, then there is an essentially unique stratification $\underline{\xi}$ of M in Q which is over $\underline{K}$. This theorem reduces the description of abstract regular neighbourhoods of M to the description of abstract stratifications $\underline{\xi}$ over $\underline{K}$ - a generalization of the description of bundles over a given base space.

We have now generalized to stratifications the first main result about block bundles. In Chapter 4 we consider transversality, proving this theorem: Given polyhedra X, Y in a manifold M. Then there is an arbitrarily small isotopy of M which carries Y "block transverse" to a stratification of X in M. Further, if a subpolyhedron Y' of Y (which satisfies a certain natural condition) is already block transverse to the stratification, then we may keep Y' fixed during the isotopy. This theorem is new, I believe, even in case X, Y are submanifolds of M and Y' of Y. The difficulties I mentioned before prevent me from proving (or disproving) that block transversality is symmetric.

As an application of the relative transversality theorem, we prove in Chapter 5 an analogue of Thom's theorem [27] which exhibited an isomorphism between the group of cobordism classes of n-dimensional differentiable manifolds, and the n-th homotopy group of the spectrum of Thom complexes of classifying bundles for the orthogonal groups. One would like to define polyhedra X and Y to be "cobordant" if there is a polyhedron W whose boundary is the disjoint union of X and Y. However, every polyhedron X without boundary would then bound the cone cX; so this cobordism theory is trivial. We define non-trivial theories thus: Let $\mathfrak{F}$ be a family of polyhedra (satisfying certain conditions). A polyhedron X is an "$\mathfrak{F}$-polyhedron" if (roughly) for every point $x \in X$, its link $\ell k(x, X) \in \mathfrak{F}$. Now "$\mathfrak{F}$-theory" is the cobordism theory in which X, Y and W are all required to be $\mathfrak{F}$-polyhedra. An "$\mathfrak{F}$-classifying" stratification $\underline{Y}/\underline{\mu}$ is a stratification of an

$\mathfrak{F}$-polyhedron in a manifold which satisfies an appropriate universal property for "morphisms" of stratified $\mathfrak{F}$-polyhedra $\underline{K} \to \underline{U}$. By such morphisms into $\underline{U}$ are classified all stratifications over $\underline{K}$ in manifolds. Hence, in the light of Chapter 3, we can classify all abstract regular neighbourhoods (Q, M), where M is a fixed polyhedron and Q a (variable) manifold. If $\underline{\xi}/\underline{K}$ is a stratification of M in Q, then the "Thom space" of $\underline{\xi}$ is defined by identifying to a point the complement of a regular neighbourhood of M in Q. Then our analogue of Thom's theorem is: For each family $\mathfrak{F}$, there is an isomorphism χ : $\mathfrak{F}$-theory $\to$ the homotopy groups of the spectrum of Thom complexes of $\mathfrak{F}$-classifying stratifications.

As another consequence of the relative transversality theorem we prove: Let μ/K be a disk block bundle (in the sense of Rourke and Sanderson) with K any polyhedron. Let M be an abstract regular neighbourhood of K which is a manifold. By extending μ over M we obtain an abstract regular neighbourhood μ^*M of μ which is a manifold. Then the function μ^* : (Manifold regular neighbourhoods of K) $\to$ (Manifold regular neighbourhoods of μ) is a bijection.

In Chapter 6 this question is raised: given $X \subseteq Y$, when does X have a (disk)-block bundle neighbourhood in Y? A necessary condition is that X be "locally flat" in Y; that is, for every point $x \in X$, there are neighbourhoods U of x in X and V of x in Y such that $U \subseteq V$, and the pair (V, U) is p.ℓ. isomorphic to the pair $(U \times D, U \times 0)$, where D is a disk, 0 an internal point of D. However, this condition is not sufficient. A primary type of obstruction is defined and, in principle, classified. These obstructions are not sufficient; the usual difficulties arise. I have tried to elucidate the problem though I cannot solve it. It is related to the difficulty in proving block transversality symmetric.

In Chapter 7 I have listed various problems concerning stratified polyhedra that interest me; some spring directly from the material of the previous chapters; some are not yet related to this material and I think they should be.

In the Appendix is outlined the theory of block bundle flags. I have used an even more general setting, in order to deal simultaneously with other structures needed in this paper. The method of developing the theory, including the statements of results and their proofs, are largely a straightforward generalization of the corresponding parts of Rourke and Sanderson's [19, I and II]. Accordingly, I have only given those details of proof which might not be immediately apparent.

As this paper was being prepared for the press, I proved that block transversality between a polyhedron and a submanifold of a manifold is symmetric. I have inserted a proof as Chapter 8.

Contents

Chapter 1. Definitions

§1. General Polyhedra

This section contains our polyhedral and combinatorial foundations. We follow Cohen [7, section 1] with a few additions from Zeeman's more general treatment [32, chapters 1 and 2].

A simplicial complex B will always be assumed locally finite and contained in some Euclidean space R^q. Hence B is finite-dimensional and countable. $|B|$ denotes the underlying topological space of B. The notation $A < B$ means: A is a subcomplex of B.

We shall denote simplexes of a simplicial complex B by Roman letters r, s, t; and if the vertices of s are $v_o, \ldots, v_a$, then we write: $s = \langle v_o, \ldots, v_a \rangle$. If $s, t \in B$, we define

$$\langle s, t \rangle = \begin{cases} \text{the simplex of } B \text{ whose vertices are} \\ \text{those of } s \text{ together with those of } t, \text{ if it exists;} \\ \emptyset, \text{ otherwise} \end{cases}$$

A barycentre of s is any point $\hat{s}$ of the interior of s. If a barycentre is given for every simplex of B, then we can form a corresponding first-derived subdivision B' of B. If $A < B$ is a subcomplex, A' will denote the induced subdivision of A. The dual cone to s in A is defined as

$$s^*A = \begin{cases} \cup\{\langle \hat{s}, \hat{r}_1, \hat{r}_2, \ldots \rangle : s, r_1, r_2, \ldots \in A \text{ and} \\ \quad s < r_1 < r_2 < \ldots\}, \text{ if } s \in A; \\ \emptyset, \text{ if } s \notin A. \end{cases}$$

A triangulation of a topological space X is a pair (B, h), where B is a simplicial complex and $h : |B| \to X$ is a homeomorphism. We shall usually not mention h, confound B with h(B), and say: B is a triangulation of X. A polyhedron consists of a topological space X and a maximal family of related triangulations of X. (Triangulations (B, h) and (A, g) of X are related if there exist subdivisions B' of B, A' of A and a simplicial isomorphism $a : B' \approx A'$ such that the composition $g \cdot a = h$.) We shall usually not mention the family of triangulations, and refer to the polyhedron simply as X; however, any triangulation of X will be assumed to come from its family of triangulations. Thus all our polyhedra are finite-dimensional, locally compact, separable, metrizable spaces. We shall occasionally assume that a polyhedron X is endowed with a metric d such that $d : X \times X \to R^1$ is piecewise linear (defined below).

Z is a subpolyhedron of X if there is a triangulation (B, h) of X and a subcomplex $A < B$ such that (A, h) is a triangulation of Z. Thus Z is always closed in X. Unless otherwise stated, the notation $Z \subseteq X$ will mean: Z is a subpolyhedron of X. More generally - still following Cohen - the notation $(X, \{X_i\})$ denotes a polyhedron X together with a family of subpolyhedra of X such that there exists a single triangulation of X in which every X_i is triangulated by a subcomplex. In practise our families will satisfy the stronger condition of being locally finite: every point $x \in X$ has a neighbourhood which meets only finitely many X_i. In case the family of subpolyhedra forms a finite, monotone sequence $X_o \subseteq X_1 \subseteq \dots \subseteq X_n = X$, we say $(X, \{X_i\})$ is a filtered polyhedron, denoted: $(X_n, \dots, X_o)$, or simply: (X).

Since filtered polyhedra occur so often in this paper, we offer some conventions for their use: Let $(X_n, \dots, X_a)$ be a filtered polyhedron. We shall always assume $X_i = \emptyset$ if $i < 0$. (X) is of length ℓ if there are at most $\ell+1$ non-empty terms in the filtration, thus $(X_n, \dots, X_a)$ is of length $n-a$ (or less). Given $b \lneq a$ and $m \geq n$, we can extend the filtration $(X_n, \dots, X_a)$ to a filtration $(X_m, \dots, X_b)$ by setting $X_m = \dots = X_n = X$ and $X_{a-1} = \dots = X_b = \emptyset$. We shall usually not distinguish between this extension and the original filtration. (They are in fact "equivalent" filtrations in the sense of §2.) If $Z \subseteq X$, then Z is filtered by the restriction of (X); $(Z) = (Z \cap X_n, \dots, Z \cap X_a)$. In particular we have $(X_k) = (X_k, \dots, X_a)$ whenever $a \leq k \leq n$. Now let $(Y) = (Y_m, \dots, Y_b)$ be another filtered polyhedron. Then $X \cup Y$ and $X \cap Y$ are filtered by $(X \cup Y)_i = X_i \cup Y_i$, $(X \cap Y)_i = X_i \cap Y_i$; here we have to extend (X) and (Y) so that i runs from $\min(a, b)$ to $\max(m, n)$. $X \times Y$ is filtered by the subpolyhedra $(X \times Y)_k = \cup \{X_i \times Y_j : i + j = k,\ a \leq i \leq n,\ b \leq j \leq m\}$. We say $(X) \subseteq (Y)$ if $X_i \subseteq Y_i$ for $i = a, \dots, n$; thus we must have $b \leq a$ and $m \geq n$. (Note that (X) is not required to be a restriction of (Y).) If $X_n \subseteq Y_b$, then Y has the filtration $(Y, X) = (Y_m, \dots, Y_b, X_n, \dots, X_a)$.

A continuous map $f : X \to Y$ between polyhedra is piece-wise linear (abbreviated to: p.ℓ.) if there are triangulations (B, h) of X and (A, g) of

Y such that $g^{-1} \circ f \circ h : B \to A$ is simplicial. Unless otherwise stated we shall assume also that f is <u>proper</u>; that is, if $Z \subseteq Y$ is compact, then so is $f^{-1}(Z)$. If $\{X_i\}$, $\{Y_j\}$ are locally finite families of subpolyhedra of X and Y and $f : X \to Y$ is p.ℓ. and proper, then we can choose B and A so that each X_i and Y_j is triangulated by a subcomplex. If the index set for $\{X_i\}$ is contained in that for $\{Y_j\}$ and $f(X_i) \subseteq Y_i$ for all i, then we write: $f : (X, \{X_i\}) \to (Y, \{Y_j\})$, or, in the case of filtered polyhedra: $f : (X) \to (Y)$.

I denotes the unit interval $[0, 1]$. If $f, g : (X, \{X_i\}) \to (Y, \{Y_j\})$ are p.ℓ. embeddings, an <u>isotopy</u> h_t of f to g is a p.ℓ. embedding $h : (X \times I, \{X_i \times I\}) \to (Y \times I, \{Y_j \times I\})$ such that for all $t \in I$ there is a p.ℓ. embedding $h_t : (X, \{X_i\}) \to (Y, \{Y_j\})$ with

$h(x, t) = (h_t(x), t)$, for all $x \in X$;

$h_0 = f$;

$h_1 = g$.

In case $(X, \{X_i\}) \subseteq (Y, \{Y_j\})$, that is, $X \subseteq Y$ and $X_i \subseteq Y_i$ for all i, then f will usually be the inclusion, and we shall then refer to h_t as an isotopy of $(X, \{X_i\})$ in $(Y, \{Y_j\})$. In case $(X, \{X_i\}) = (Y, \{Y_i\})$, we shall usually assume $h(X \times I) = Y \times I$ and $h(X_i \times I) = Y_i \times I$ for all i.

Given $(Y, \{Y_j\})$ and X, $Z \subseteq Y$. Then $(V, \{V_j\})$ is a <u>regular neighbourhood</u> of X rel Z in $(Y, \{Y_j\})$ with <u>frontier</u> fr V if there exist a simplicial triangulation B of Y in which X, Z and each Y_j are triangulated by full subcomplexes A, C and B_j, and a first derived subdivision B' of B such that:

$V = |N(X-Z, B')|$;

$V_j = |N(X-Z, B'_j)|$ for all j;

$\text{fr } V = |\ell k(X-Z, B')|$.

In case (Y) is a filtered polyhedron, (V) has the restriction filtration, and so does $(\text{fr } V)$. In some contexts we shall also use $(V^{\bullet})$ for the frontier of V. $(Y, \{Y_j\})$ is a <u>weak</u> regular neighbourhood of X rel Z with frontier fr Y if it is a regular neighbourhood in $(\text{fr } Y \times I \cup Y, \{(Y_j \cap \text{fr } Y) \times I \cup Y_j\})$ where $\text{fr } Y \times I \cup Y$ is formed from the disjoint union by identifying $\text{fr } Y \times 0$ to fr Y in the obvious way.

We quote Cohen's Uniqueness Theorem for regular neighbourhoods [7, Addendum 3.3]:

<u>Theorem 1.1.</u> Let $(V, \{V_j\})$ and $(W, \{W_j\})$ be regular neighbourhoods of X rel Z in $(Y, \{Y_j\})$. Then there is an isotopy h_t of $(Y, \{Y_j\})$ such that: $h_1 V = W$, $h_1 V_j = W_j$ for all j, $h_1 \text{ fr } V = \text{fr } W$; h_t is the identity on $X \cup Z$.

Cohen's theorem does not include the statement about frontiers, but it is easily inferred from his proof.

Let y be a point of Y. Then $st(y; Y, \{Y_j\})$ denotes a regular neighbourhood of y in $(Y, \{Y_j\})$, and $\ell k(y; Y, \{Y_j\})$ its frontier. In case (Y) is a filtered polyhedron, we write: $st(y; Y_n, \ldots, Y_0)$, or: $st(y; (Y))$.

Let X be a polyhedron. A subset $\mathfrak{Z}$ of X is a <u>sub-polyspace</u> of X if both its closure in X, $\text{cl}\,\mathfrak{Z}$, and $\text{cl}\,\mathfrak{Z} - \mathfrak{Z}$ are subpolyhedra of X. This definition is less general than Zeeman's (see [32]), but will suffice for our purposes. $\mathfrak{Z}$ is then a polyhedron in its own right, but not usually a subpolyhedron of X. (For example, the inclusion $\mathcal{Z} \to X$ is not usually proper). However $\mathcal{Z}$ is "locally" a subpolyhedron of X; that is, if x is a point of $\mathcal{Z}$, then $st(x, \mathcal{Z}) = st(x, cl\,\mathcal{Z})$ for sufficiently small stars, so if we choose a sufficiently small $st(x; X, cl\,\mathcal{Z})$ we have that $st(x, \mathcal{Z})$ is a subpolyhedron of $st(x, X)$. In hopes of reducing confusion I shall use script Roman letters to denote sub-polyspaces.

Given $(Y, \{Y_j\})$ and $Y^* \subseteq Y$. A <u>collar neighbourhood</u> of Y^* in $(Y, \{Y_j\})$ is a regular neighbourhood $(V, \{V_j\})$ of Y^* in $(Y, \{Y_j\})$ together with a p.ℓ. isomorphism $p : Y^* \times I \to V$ which corresponds

$Y_j^* \times I \longleftrightarrow V_j$ for all j, where $Y_j^* = Y^* \cap Y_j$;

$Y^* \times 0 \longleftrightarrow Y^*$ in the obvious way;

$Y^* \times 1 \longleftrightarrow \operatorname{fr} V$.

If Y^* has a collar neighbourhood in $(Y, \{Y_j\})$, then we write:

$(Y^*, \{Y_j^*\}) < \operatorname{bdy}(Y, \{Y_j\})$, or simply: $Y^* < \operatorname{bdy}(Y, \{Y_j\})$.

If also $X \subseteq Y$ and $(Y^*, \{X \cap Y^*, Y_j^*\}) < \operatorname{bdy}(Y, \{X, Y_j\})$, then a regular neighbourhood $(V, \{V_j\})$ of X in $(Y, \{Y_j\})$ meets Y^* <u>regularly</u> if $(V, \{V \cap Y^*, V_j\})$ is a regular neighbourhood of X in $(Y, \{Y^*, Y_j\})$. It follows that $(Y^*, \{Y^* \cap \text{everything}\} < \operatorname{bdy}(Y, \{\text{everything}\})$ (where "everything" refers to the Y_j, the V_j, X and fr V).

Y^* is <u>locally collared</u> in $(Y, \{Y_j\})$ if for every point $y \in Y^*$, when we take $st(y; Y, \{Y^*, Y_j\})$, then its intersection with Y^*, namely $st(y; Y^*)$, has a collar neighbourhood in $st(y; Y, \{Y_j\})$.

Proposition 1.2. 1. Given $(Y, \{Y_j\})$. Then $Y^* \subseteq Y$ is locally collared in $(Y, \{Y_j\})$ if and only if it is collared in $(Y, \{Y_j\})$.

2. If $p : Y^* \times I \to V$ and $q : Y^* \times I \to W$ are collar neighbourhoods of Y^* in $(Y, \{Y_j\})$, then there is an isotopy h_t of $(Y, \{Y_j\})$ such that:

$h_1 \circ p = q$;

h_t is the identity on Y^*.

3. Given $X^*, Y^* < \mathrm{bdy}(Y, \{Y_j\})$ such that if $Z = X^* \cap Y^*$, then $Z < \mathrm{bdy}(X^*, \{X_j^*\})$ and $Z < \mathrm{bdy}(Y^*, \{Y_j^*\})$. Then $X^* \cup Y^* < \mathrm{bdy}(Y, \{Y_j\})$.

We shall need a generalization of Cohen's Stellar Neighbourhood Theorem [7, Theorem 6.1 (c)]:

Theorem 1.3. Given $(Y, \{Y_j\})$ and $X, Z \subseteq Y$. Let $(V, \{V_j\}) \subseteq (Y, \{Y_j\})$ and $\mathrm{fr}\, V \subseteq V$ be such that there is a triangulation B of Y in which X, Z and each Y_j are covered by full subcomplexes A, C and B_j and in which:

$V = |N(X-Z, B)|$;

$V_j = |N(X-Z, B_j)|$ for all j;

$\mathrm{fr}\, V = |\ell k(X-Z, B)|$;

$\mathrm{fr}\, V$ is collared in $(V, \{V_j\})$ and in $(\mathrm{cl}[Y-V], \{\mathrm{cl}[Y_j-V]\})$.

Then $(V, \{V_j\})$ is a regular neighbourhood of X rel Z in $(Y, \{Y_j\})$ with frontier $\mathrm{fr}\, V$.

Cohen's theorem does not include the family $\{Y_j\}$, but his proof applies also to this situation.

Proposition 1.4. Given $(Y, \{Y_j\})$ and $Z \subseteq X \subseteq Y$, and a regular neighbourhood $(U, \{W, U_j\})$ of Z in $(Y, \{X, Y_j\})$. Set $Y' = \mathrm{cl}[Y-U]$,

$X' = X \cap Y'$, $Y'_j = Y_j \cap Y'$. Consider a subpolyhedron $(V, \{V_j\}) \subseteq (Y', \{Y'_j\})$. Then the following conditions on V are equivalent:

1. $(V, \{V_j\})$ is a regular neighbourhood of X' in $(Y', \{Y'_j\})$ which meets fr U regularly;

2. $(U \cup V, \{U_j \cup V_j\})$ is a regular neighbourhood of X in $(Y, \{Y_j\})$, and $(V \cap \text{fr } U, \{V_j \cap \text{fr } V\})$ is a regular neighbourhood of fr W in $(\text{fr } U, \{Y_j \cap \text{fr } U\})$ with frontier "fr V" $\cap$ fr U, where "fr V" is defined to be $V \cap \text{cl}[Y' - V]$. (See diagram 1, p. 1.9).

The proof is a straight-forward application of the preceding results.

Given $Z \subseteq X$, then we define $X \times_Z I$ to be a regular neighbourhood of $X \times 0$ rel $Z \times 0$ in $X \times [0, 2]$. Roughly speaking we may think of $X \times_Z I$ as obtained from $X \times I$ by identifying $z \times I$ to $z \times 0$ for every point $z \in Z$. With this in mind we identify X with $X \times 0 \subseteq X \times_Z I$ and write $X \times_Z 1$ for fr $X \times_Z I$. Here is a restatement of Cohen's [7, Proposition 7.5]:

Proposition 1.5. Let $(Y, \{Y_j\})$ be a weak regular neighbourhood of X rel Z (so that $Z \subseteq X$), and let $(V, \{V_j\})$ be a regular neighbourhood of X rel Z in $(Y, \{Y_j\})$. Then $Z \subseteq \text{fr } V$ and there is a p.ℓ. isomorphism

$p : \text{cl}[Y-V] \longrightarrow \text{fr } V \times_Z I$ which corresponds

$\text{cl}[Y_j - V] \longleftrightarrow (V_j \cap \text{fr } V) \times_Z I$ for all j ;

fr V $\longleftrightarrow$ fr $V \times 0$ in the obvious way;

fr Y $\longleftrightarrow$ fr $V \times_Z 1$.

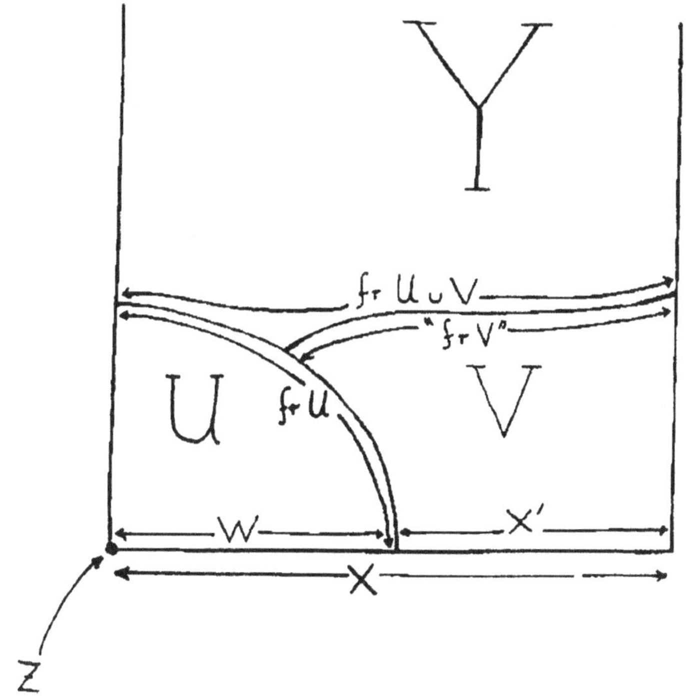

diagram 1

Again, we have generalized Cohen's result to deal with a given family of subpolyhedra.

Given a filtered polyhedron $(Y_q, \ldots, Y_o)$. Then $bdy^o Y$ is defined as the set of points $y \in Y$ such that there exist compact polyhedra $(F_n, \ldots, F_o)$ and a p.ℓ. isomorphism $p : st(y, (Y)) \longrightarrow D \times c(F)$, which corresponds $y \longleftrightarrow v \times c$, where D is a disk of dimension ≥ 1, $v \in bdy\ D$. $bdy^o Y$ is a poly-subspace, but not always a subpolyhedron, of Y; for example, if Y is the region of the plane bounded by a figure 8, then $bdy^o Y$ is not closed in Y. So we take the closure, and define $bdy\ Y = cl[bdy^o Y]$.

The reader should bear in mind that, though $bdy(bdy^o X) = bdy^o(bdy^o X) = \emptyset$, $bdy(bdy\ X)$ need not be empty. For example, in Euclidean 3-space, R^3, take $X = R^2 \cup \{(x, y, z) : x = 0,\ 0 \leq y, z \leq 1\}$; then $bdy\ X = \{(0, y, 0), (0, y, 1), (0, 0, z) : 0 \leq y, z \leq 1\}$, and $bdy(bdy\ X) = \{(0, 0, 0), (0, 1, 0)\}$.

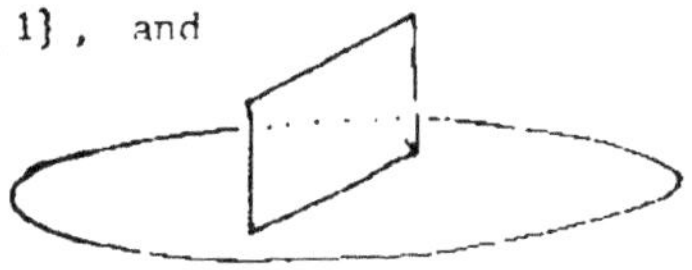

Now say $X \subseteq Y_o$ and $(X_n, \ldots, X_o)$ is a filtration of X. Then we have $bdy(Y, X) \subseteq Y$; we set $bdy_{(Y)}(X) = X \cap bdy(Y, X)$. If the filtrations of Y and X are clear in context, we shall write: $bdy_Y X$. If further we are given $Z \subseteq X$, then the notation $Z < bdy_Y X$ will mean: there is some $W \subseteq Y$ with $W < bdy(Y, X)$ such that $Z = W \cap X$.

If (and only if) F and G are compact polyhedra, then their <u>join</u> $F*G$ is a polyhedron. (It is defined in Cohen's [7, section 1].) If $F = \emptyset$, the empty set, then $F*G = G$. If F is a single point, then $F*G$ is the <u>cone</u>

on G, and will usually be denoted cG. G is thus identified with the base of the cone; c is its vertex. If $G' \subseteq G$, then cG' is naturally contained in cG, and is called a subcone of cG.

Let M be an n-manifold, $N \subseteq$ bdy M an (n-1)-submanifold, possibly with boundary; or N may be empty. A polyhedron which is p.ℓ. isomorphic to M-N for some choice of M and N will be called an open manifold. Note that M-N is a sub-polyspace of M.

A cell complex K is a countable, locally finite set of (p.ℓ) disks contained in some polyhedron, such that for all $\sigma, \tau \in K$:

$\sigma \cap \tau$ is a union of cells of K;

if $\sigma \neq \tau$, then the interiors of σ and τ are disjoint;

the boundary of σ is a union of cells of K.

Thus $\cup\{\sigma : \sigma \in K\}$ is a polyhedron, denoted $|K|$. τ is a face of σ if $\tau \subseteq \sigma$; we write: $\tau < \sigma$. The notation $L < K$ will mean: L is a subcomplex of K; that is, L is a cell complex and $L \subseteq K$. For every $\sigma \in K$ we have the subcomplex $\langle\sigma\rangle$ defined as $\{\tau \in K : \tau < \sigma\}$. A standard ordering of a cell complex K is a denumeration of its cells, $\sigma_1, \ldots, \sigma_r, \ldots$ such that if $\tau < \sigma$ then τ precedes σ in the denumeration. (If K is finite one can just list all the 0-cells, then all the 1-cells, and so on.)

The letter ϵ will be used in the sense of "positive and arbitrarily small", and δ in the sense of "positive and sufficiently small". For example: given ϵ, there is a rational number δ which is less than ϵ.

I use the verb "to cover" meaning: to have the same underlying space as". For example, if B and C are triangulations of a polyhedron X, then B covers C.

§2. Disjunctions and Varieties

A disjunction of a polyhedron X is a family $\{\mathfrak{X}_o, \ldots, \mathfrak{X}_n\}$ of sub-polyspaces (not usually subpolyhedra) of X such that: X is the disjoint union of the $\mathfrak{X}_i$'s; and for each i, $\mathrm{cl}\, \mathfrak{X}_i$ is a union of $\mathfrak{X}_j$'s.

We write $\mathfrak{X}_j < \mathfrak{X}_i$ if $\mathfrak{X}_j \subseteq \mathrm{cl}\, \mathfrak{X}_i$ and $\mathfrak{X}_j \neq \mathfrak{X}_i$. This defines a partial ordering on $\{\mathfrak{X}_i\}$; and one can renumber them so that if $\mathfrak{X}_j < \mathfrak{X}_i$, then $j < i$. This condition we shall henceforth assume.

There is a filtration of X associated to $\{\mathfrak{X}_i\}$, namely: $X_i = \cup\{\mathfrak{X}_j : j \leq i\}$. Conversely, given any filtration $(X) = (X_n, \ldots, X_o)$ of X, there is an associated disjunction of X, namely: $\mathfrak{X}_i = X_i - X_{i-1}$.

A second disjunction $\{\mathfrak{X}'_j\} = \{\mathfrak{X}'_1, \ldots, \mathfrak{X}'_m\}$ of X refines $\{\mathfrak{X}_i\}$ if each component of each $\mathfrak{X}'_j$ is contained in some $\mathfrak{X}_i$. If also $\{\mathfrak{X}\}$ refines $\{\mathfrak{X}'\}$, then we say $\{\mathfrak{X}\}$ and $\{\mathfrak{X}'\}$ are equivalent. The corresponding definitions for filtrations of X are made via the correspondence between filtrations and disjunctions just described. The advantage of using disjunctions is this: if $\{\mathfrak{X}'\}$ is a re-ordering of $\{\mathfrak{X}\}$, then $\{\mathfrak{X}'\}$ and $\{\mathfrak{X}\}$ are clearly equivalent; but the relationship between their associated filtrations is not so clear.

For most of the theorems (though not the definitions) we shall need a special type of disjunction: Let $(Y_q, \ldots, Y_o)$ be a filtered polyhedron, and let $X \subseteq Y_o$ be given. A disjunction $\{\mathfrak{X}_o, \ldots, \mathfrak{X}_n\}$ of X is a variety of X in (Y) if: whenever x is a point of $\mathfrak{X}_i$, then there exist a p.ℓ. disk D, a compact filtered polyhedron $(G, F) = (G_q, \ldots, G_o, F_n, \ldots, F_{i+1}, \emptyset)$ and a p.ℓ. isomorphism

$h : st(x; Y_q, \ldots, Y_o, X_n, \ldots, X_i) \longrightarrow D \times c(G, F)$

with $x \longleftrightarrow v \times c$, for some $v \in D$.

Here $(X_n, \ldots, X_o)$ is the filtration associated to $\{\mathfrak{X}_i\}$. This definition depends only on the equivalence class of $\{\mathfrak{X}_i\}$, since one only needs the equivalence class of the filtration (G, F). The filtration of X associated to a variety of X in (Y) will be called a variety filtration of X in (Y).

The most important varieties for our purposes occur this way: Given polyhedra $X \subseteq Y$ and a point x in X, we define the intrinsic dimension of x in (Y, X), denoted $d(x; Y, X)$ by: $d(x; Y, X) \geq i$ if there are compact polyhedra $G \geq F$, an i-disk D^i, and a p.ℓ. isomorphism

$h : st(x; Y, X) \longrightarrow D^i \times c(G, F)$

with $x \longleftrightarrow v \times c$, for some $v \in D^i$.

$d(x; Y, X) = i$ if it is $\geq i$ but $\not\geq i+1$. The intrinsic variety of X in Y is defined by: $\mathfrak{X}_i = \{x \in X : d(x; Y, X) = i\}$. It will be proved in Lemma 1.10 that the intrinsic variety is indeed a variety. For now we note that if X is of dimension n (and all our polyhedra are finite-dimensional), then X is the disjoint union of $\mathfrak{X}_o, \ldots, \mathfrak{X}_n$. One can define $\mathfrak{X}_i$ directly as the set of x in X such that $st(x; Y, X) \underset{p.\ell.}{\approx} D^i \times c(G, F)$ as above, where the pair (G, F) is neither the suspension of, nor the cone on, another pair. This pair (G, F) will be called the basic link of x in (Y, X), denoted $b.\ell k(x; Y, X)$. The associated filtration to $\{\mathfrak{X}_i\}$ is the intrinsic filtration of X in Y.

By setting $Y = X$ we obtain definitions of the intrinsic variety and filtration of X, and the intrinsic dimension and basic link of a point x in

X. The intrinsic variety of X in Y refines the intrinsic variety of X.

Akin [1], following Armstrong and Zeeman [4], has defined the intrinsic dimension of x in (Y, X) slightly differently. His $d'(x; Y, X) \geq i$ if (to use our previous notation) $h : st(x; Y, X) \longrightarrow D^i \times c(G, F)$ with $x \longleftrightarrow v \times c$, where v is in the interior of D^i. He defines the intrinsic skeleta of X in Y by: $I^i(Y, X) = \{x \in X : d'(x; Y, X) \leq i\}$. We shall refer to the disjunction of X by $\mathfrak{X}_i' = \{x \in X : d'(x; Y, X) = i\}$ as the skeletal variety of X in Y; from it we derive the skeletal filtration of X in Y. It will be proved in Lemma 1.10 that the skeletal variety is indeed a variety. By setting Y = X we obtain the skeletal variety of X, etc.

The skeletal variety of X in Y refines the intrinsic variety of X in Y. In fact, it will follow from Lemma 1.10 that every variety of X in Y refines the intrinsic one. They are not always equal, as these examples show:

1. Let M be a manifold with non-empty boundary. Then its skeletal variety is {M - bdy M, bdy M}; but its intrinsic variety is just {M}.
2. Let $\{\mathfrak{X}_i\}$, $\{\mathfrak{X}_j'\}$ be the intrinsic and skeletal varieties of X. Then the intrinsic variety of X x I is $\{\mathfrak{X}_i \times I\}$; but the skeletal variety is $(\mathfrak{X} \times \mathfrak{I})_i = \mathfrak{X}_i \times \{0, 1\} \cup \mathfrak{X}_{i-1} \times (0, 1)$.
3. If $G \geq F$ are compact and $\{\mathcal{J}_i\} = \{\mathcal{J}_n, \ldots, \mathcal{J}_k\}$ is a variety of F in G, then $\{c, (c\mathcal{J}_i - c)\}$ is a variety of cF in cG. If $\{\mathcal{J}_i\}$ is the intrinsic variety of F in G, then $\{c, (c\mathcal{J}_i - c)\}$ is the intrinsic variety if cF in cG - unless (G, F) is a cone-pair or a suspension pair, in which case it is

$\{c\mathcal{J}_k, (c\mathcal{J}_i - c) : i = k+1, \ldots, n\}$. If $\{\mathcal{J}'_n, \ldots, \mathcal{J}'_\ell\}$ is the skeletal variety of F in G, then the skeletal variety of cF in cG is $\{c \cup \mathcal{J}'_\ell, (c\mathcal{J}'_\ell - c) \cup \mathcal{J}'_{\ell+1}, \ldots, (c\mathcal{J}'_{n-1} - c) \cup \mathcal{J}'_n, c\mathcal{J}'_n - c\}$ - unless (G, F) is a suspension pair, in which case it is $\{\mathcal{J}'_\ell, c\mathcal{J}'_\ell \cup \mathcal{J}'_{\ell+1}, \ldots, (c\mathcal{J}'_{n-1} - c) \cup \mathcal{J}'_n, c\mathcal{J}'_n - c\}$.

4. Let $(X_n, \ldots, X_o)$ be a variety filtration of X in Y. Then for each $k = 0, \ldots, n$, $(X_k, \ldots, X_o)$ is associated to a variety of X_k in Y. However, even if (X) is intrinsic, (X_k) need not be; for example, the intrinsic filtration of the one-point union $S^1 \vee S^2$ is $(S^1 \vee S^2, S^1$, base point), but (S^1) and not $(S^1$, base point) is the intrinsic filtration of S^1.

5. Whitney [29] has defined a "stratification" of the analytic variety V as an expression of V as the disjoint union of a locally finite set of analytic manifolds, each of constant dimension, called the "strata", such that the frontier of each stratum is the union of a set of lower dimensional strata. I have used the term "variety" in our analogous p. ℓ context for two reasons: first, I wish a p. ℓ. "stratification" to be analogous to Thom's definition [28] of an "ensemble stratifié", which has more structure than Whitney's stratification; second, an analytic variety has more structure than just its underlying topological space, so a "p. ℓ variety" may reasonably denote a polyhedron together with additional structure. The connection between Whitney stratifications and p. ℓ. varieties is provided by Łojasiewicz [14]. One can put on a stratified analytic variety a canonical p. ℓ structure in which the closure of every stratum is a subpolyhedron. In this p. ℓ structure the analytic strata form a p. ℓ. variety.

6. Even when there is a canonical "simplest" analytic Whitney stratification, the resulting p.ℓ. variety may not be intrinsic. Let us consider the Brieskorn varieties; say V, of real dimension 10, in complex space C^6 (which we regard as real space R^{12}), is defined by:
$z_0^2 + z_1^2 + \dots + z_4^2 + z_5^3 = 0$. The only singular point of V is the origin, c. So V has the analytic stratification (V-c, c). The intersection of (R^{12}, V^{10}) with the unit sphere in R^{12} is a pair (S^{11}, Σ^9), and it is known that Σ^9 is an exotic homotopy sphere (see Milnor [16]). So in the p.ℓ. category, Σ^9 is simply a 9-sphere, V is a p.ℓ. 10 manifold, and its intrinsic variety is just V. However, the Σ^9 is knotted in S^{11} (see [16]) smoothly and piecewise linearly; hence the intrinsic variety of V in R^{12} is (V-c, c). On the other hand, this resource fails us if we include $C^6 \subseteq C^7$; the intrinsic variety of V^{10} in R^{14} is just V, since local knotting cannot occur piecewise linearly in this codimension; whereas the analytic stratification is still (V-c, c). This example explains why we work with varieties in general, and not just intrinsic (or skeletal) ones.

7. A subsidiary reason for using general varieties is this: One of our main questions is, given a polyhedron X, how many regular neighbourhoods M does it have which are manifolds (up to p.ℓ. isomorphism rel X)? We shall only be able to discuss the weaker question, given X and a variety of X, how many M are there such that the intrinsic variety of X in M is the given one? If dim M - dim X $\geq$ 3 (and under another simple hypothesis), the intrinsic variety of X in M equals the intrinsic variety of X. I shall

often, I admit, be compelled to make these assumptions. But at least the reader may find it helpful to think of a general variety of a polyhedron X as its intrinsic variety plus other manifolds along which we allow local knotting when embedding X in manifolds.

Theorem 1.6. Given $(Y_q, \ldots, Y_o)$, $X \subseteq Y_o$ and a variety filtration $(X_n, \ldots, X_o)$ of X in (Y). Then:

1. Each $\mathfrak{X}_i$ of the associated variety is an open manifold (see §1 for the definition);

2. If K is a cell-complex triangulation of Y in which each Y_j or X_i is covered by a subcomplex K_j or L_i, then $\mathrm{bdy}(Y, X)$ and $\mathrm{bdy}_Y X$ are covered by subcomplexes $\mathrm{bdy}(K, L)$ and $\mathrm{bdy}_K L$ of K.

3. Let f_t be an isotopy of (X). Then f_t extends to an isotopy of (Y, X).

4. If in fact (X) is the intrinsic filtration of X in (Y) and K is a cell complex triangulation of (Y) in which X is covered by a subcomplex, then each X_i is covered by a subcomplex of K.

These results are due to Akin [1]. His work deals with the skeletal filtration of X in Y, but the proofs apply to the present situation. By virtue of 1, if $\{\mathfrak{X}_i\}$ is a variety of X, we shall henceforth speak of $\mathfrak{X}_i$ as a manifold of the variety.

Lemma 1.7. Let $(F, \{F_i\})$ and $(F', \{F'_i\})$ be compact polyhedra with families of subpolyhedra indexed by the same set Ω. Assume that for some $\hat{\imath} \in \Omega$, $F_{\hat{\imath}} = \emptyset$. Given a p.$\ell$. isomorphism $f : D^n \times c'(F', \{F'_i\}) \to D^k \times c(F, \{F_i\})$, where D^n is an n-disk, D^k a k-disk. Then $n \leq k$, $F'_{\hat{\imath}}$ is either S^{k-n-1} or

D^{k-n-1} (here $S^{-1} = D^{-1} = \emptyset$), and there is a p.ℓ. isomorphism $g : (F', \{F'_i\}) \longrightarrow F'_{\hat{i}} * (F, \{F_i\})$.

The proof is by induction on n. If $n = 0$, we have

$$\begin{aligned}(F', \{F'_i\}) &\underset{p.\ell.}{\approx} \ell k(c', c'(F', \{F'_i\})) \\ &\underset{f}{\longrightarrow} \ell k(fc', D^k \times c(F, \{F_i\})) \\ &\underset{p.\ell.}{\approx} \ell k(fc', D^k \times c) * (F, \{F_i\})\end{aligned}$$

since $fc' \in D^k \times cF_{\hat{i}} = D^k \times c$. Under this triple composite, $F'_{\hat{i}} \underset{p.\ell.}{\approx} \ell k(fc', D^k \times c) * F_{\hat{i}} = \ell k(fc', D^k \times c)$ which is S^{k-1} or D^{k-1} according as $fc' \in \text{int}(D^k \times c)$ or $fc' \in \text{bdy}(D^k \times c)$. Thus the lemma holds for $n = 0$. For the general step, take a point $x \in \text{bdy}\, D^n$; then $x \times c' \in \text{bdy}(D^n \times cF'_{\hat{i}})$, so $f(x \times c') \in \text{bdy}\, D^k \times c$. Hence

$$\begin{aligned}D^{n-1} \times c'(F', \{F'_i\}) &\underset{p.\ell.}{\approx} \ell k(x \times c', D^n \times c'(F', \{F'_i\})) \\ &\underset{f}{\longrightarrow} \ell k(f(x \times c'), D^k \times c(F, \{F_i\})) \\ &\underset{p.\ell.}{\approx} D^{k-1} \times c(F, \{F_i\}).\end{aligned}$$

Applying inductive hypothesis to this triple composite, we see that the lemma holds in this case, and hence for all n. Observe that $F'_{\hat{i}}$ is a sphere or a disk according as $f(y \times c') \in \text{int}(D^k \times c)$ or $\in \text{bdy}(D^k \times c)$, where y is any point of $\text{int}\, D^n$.

Lemma 1.8. Given $(Y_q, \ldots, Y_o)$, $X \subseteq Y_o$ and a filtration $(X_n, \ldots, X_o)$ of X. Let D be a disk. Then (X) is a variety filtration of X in (Y) if and only if $(D \times X) = (D \times X_n, \ldots, D \times X_o)$ is a variety filtration of $D \times X$ in $(D \times Y)$.

Let $(D \times X)$ be a variety filtration of $D \times X$ in $(D \times Y)$. Take points $r \in D$, $x \in \mathfrak{X}_i$. Then there exist compact polyhedra $(G_q, \ldots, G_o, F_n, \ldots, F_{i+1}, \emptyset)$, and a disk D'' such that

$st((r, x); D \times Y_q, \ldots, D \times X_i) \underset{p.\ell.}{\approx} D'' \times c(G, F)$. On the other hand, $st((r, x), D \times (Y, X)) \underset{p.\ell.}{\approx} st(r, D) \times st(x; (Y, X))$; and $st(r, D)$ is a disk D', while $st(x; (Y, X))$ is of the form $c(G'_q, \ldots, G'_o, F'_n, \ldots, F'_{i+1}, F'_i)$, being the cone on $\ell k(x; (Y, X))$. Thus we have $D'' \times c(G, F) \underset{p.\ell.}{\approx} D' \times c(G', F')$. Using Lemma 1.7, it follows that $c(G', F') \underset{p.\ell.}{\approx} D^* \times c(G, F)$, for some disk D^*. That is, (X) is a variety filtration of X in (Y) "near x". Since this holds for all $x \in X$, (X) is a variety filtration of X in (Y). The converse argument is easier, and we leave it to the reader.

Lemma 1.9. Given $(Y_q, \ldots, Y_o)$, and $X' \subseteq X \subseteq Y_o$. Let $(X_n, \ldots, X_o)$ be a variety filtration of X in (Y), and (W, V) a regular neighbourhhood of X' in (Y, X). Then (V) is a variety filtration of X' in (W).

If x is a point of $V - fr\, V$, then $st(x; W, V)$ is also a $st(x; Y, X)$, so a suitable product structure $D \times c(G, F)$ on $st(x; Y, X)$ gives one on $st(x; W, V)$. If $x \in fr\, V$, we use the fact that $fr(W, V)$ is bicollared in (Y, X) and Lemma 1.8 to show that $fr(V)$ is a variety filtration of $fr\, V$ in $fr(W)$. Hence there is a structure $st(x; fr(W, V)) \underset{p.\ell.}{\approx} D^k \times c(G, F)$.

Now
$$\begin{aligned} st(x; (W, V)) &\underset{p.\ell.}{\approx} st(x; fr(W, V)) \times I \\ &\underset{p.\ell.}{\approx} D^k \times c(G, F) \times I \\ &\underset{p.\ell.}{\approx} D^{k+1} \times c(G, F), \end{aligned}$$
and this is a suitable product structure. This proves Lemma 1.9.

Lemma 1.10. 1. If $(X_n, \ldots, X_o)$ is the intrinsic filtration of X in Y, then the intrinsic filtration of $D^k \times X$ in $D^k \times Y$ is $(D^k \times X_n, \ldots, D^k \times X_o)$.

2. The intrinsic filtration of X in Y is a variety filtration of X in Y.

3. The skeletal filtration of X in Y is a variety filtration of X in Y.

We have the proof of 1. to the reader.

Proof of 2.: We use induction on the dimension of X. The result is trivial if dim X = 0. For the general case, take a point $x \in X_i$; then there are compact polyhedra $G \supseteq F$ such that $st(x; Y, X) \underset{p.\ell.}{\approx} D^i \times c(G, F)$. To say $x \in X_i - X_{i-1}$ is equivalent to saying that (G, F) is not a cone-pair or a suspension-pair, by Lemma 1.7. Now $\dim F \lneq \dim X$. So by inductive hypothesis, the intrinsic filtration $(F_{n-1}, \ldots, F_o)$ of F in G is a variety filtration. By Remark 3, p. 1.15, $c(F_{n-1}, \ldots, F_o)$ is the intrinsic filtration of cF in cG; the proof of Lemma 1.9 shows that $c(F_{n-1}, \ldots, F_o)$ is also a variety filtration of cF in cG. By 1. and Lemma 1.8, $D^i \times c(F)$ is the intrinsic filtration of $D^i \times cF$ in $D^i \times cG$, and a variety filtration. But $st(x; X_n, \ldots, X_o)$ agrees with the intrinsic filtration of $st(x, X)$ in $st(x, Y)$ at least in a neighbourhood of x. This suffices to prove 2.

Proof of 3.: If $F \subseteq G$ are compact polyhedra and $(F'_{n-1}, \ldots, F'_o)$ are intrinsic skeleta of F in G, then as in 2. one shows that the intrinsic skeleta of int $D^i \times (cF - F)$ in int $D^i \times (cG - G)$ are the int $D^i \times (cF'_j - F'_j)$. The inductive proof of 2. can now be applied.

§3. Regular Neighbourhood Systems

Given a filtered polyhedron $(X) = (X_q, \ldots, X_n, \ldots, X_o)$ and $X^* < \text{bdy}(X)$. A normal (q,n)-regular neighbourhood system (abbreviated to: normal(q,n)-rns) for X_n in (X) with respect to X^* is a family of subpolyhedra of X

$\underline{N} = \{N_{j,i} : 0 \leq i \leq n,\ i \leq j \leq q\}$, defined by induction on the length n of the filtration (X_n):

ns 1 $\underline{N}$ is the disjoint union of a filtered polyhedron $(N_{,o})$ and a normal $(q-1, n-1)$-rns $\underline{N}'$;

ns 2 $(N_{,o}) = (N_{q,o}, \ldots, N_{o,o})$ is a regular neighbourhood of X_o in (X) with frontier $(\dot{N}_{,o})$ which meets X^* regularly;

ns 3 Set $X' = \text{cl}[X - N_{q,o}]$ and $X'^* = (X^* \cap X') \cup \dot{N}_{q,o}$. Then $\underline{N}'$ is a normal $(q-1, n-1)$-rns for X'_n in (X') (using the restriction filtration) with respect to X'^*. (It follows from Proposition 1.2 and Theorem 1.3 that X'^* is indeed $< \text{bdy}(X')$.)

See diagrams 2, 3, pp. 1.23, 1.24. We also speak of a normal (q,n)-rns for a disjunction $\{\mathfrak{X}_q, \ldots, \mathfrak{X}_n, \ldots, \mathfrak{X}_o\}$ of a polyhedron X with respect to X^* by passing to the filtration of X associated to $\{\mathfrak{X}_i\}$.

Given a second filtered polyhedron $(X') = (X'_q, \ldots, X'_o)$, and a normal (q,n)-rns $\underline{N}'$ for X'_n in (X'), then an isomorphism $f : \underline{N} \approx \underline{N}'$ is a p.ℓ. isomorphism $f : (X) \to (X')$ such that:

$f : N_{j,i} \underset{p.\ell.}{\approx} N'_{j,i}$ for all $N_{j,i} \in \underline{N}$;

$f : (\dot{N}_{,i}) \underset{p.\ell.}{\approx} (\dot{N}'_{,i})$ for $i = 0, \ldots, n$.

Remarks. 1. Set $\underline{N}^* = \{N_{j,i} \cap X^* : N_{j,i} \in \underline{N}\}$; then $\underline{N}^*$ is a normal (q,n)-rns for (X_n^*) in (X^*).

2. The reader may be pleased, or exasperated, to know that in the main applications of the machinery we are setting up, q will be either n or $n+1$. I have left the general q in the notation partly to get two definitions for the price of one; partly in hopes the general notion may prove useful to someone; partly from this consideration: Let us start with a filtration $(X_n, \ldots, X_o)$ of X, and an n-rns $\underline{N}$ of (X) - thus $q = n$. If we concentrate on X_{n-1}, then $\underline{N}$ includes a normal $(n, n-1)$-rns $(n-1)\underline{N}$ of (X_{n-1}) in X, namely: $(n-1)\underline{N} = \{N_{j,i} : 0 \leq i \leq n-1\}$ - thus the case $q = n+1$ is inferrable from the case $q = n$. But if we concentrate on some X_k, $k \leq n-2$, then rather than a normal $(k+1, k)$-rns $\underline{N}' = \{N_{j,i} : 0 \leq i \leq k$ and either $j \leq k$ or $j = n\}$ of (X_k) in X, it is more natural to take the normal (n, k)-rns $(k)\underline{N} = \{N_{j,i} : 0 \leq i \leq k,\ i \leq j \leq n\}$ for (X_k) in $(X_n, \ldots, X_k)$.

Proposition 1.11. 1. Given (X) and X^* as above. Then there exists a normal (q,n)-rns $\underline{N}$ for (X_n) in (X) with respect to X^*.

2. If $\underline{N}'$ is another such normal (q,n)-rns and $\underline{N}^* = \underline{N}'^*$, then there is an isotopy f_t of (X) such that:

$f_1 : \underline{N} \approx \underline{N}'$;

f_t is the identity on X^* and on X_o.

Proof of 1.: Let B be a simplicial triangulation of X such that X^* and each X_j is covered by a full subcomplex B^* and B_j of B. Let B' be a first derived subdivision of B. Set

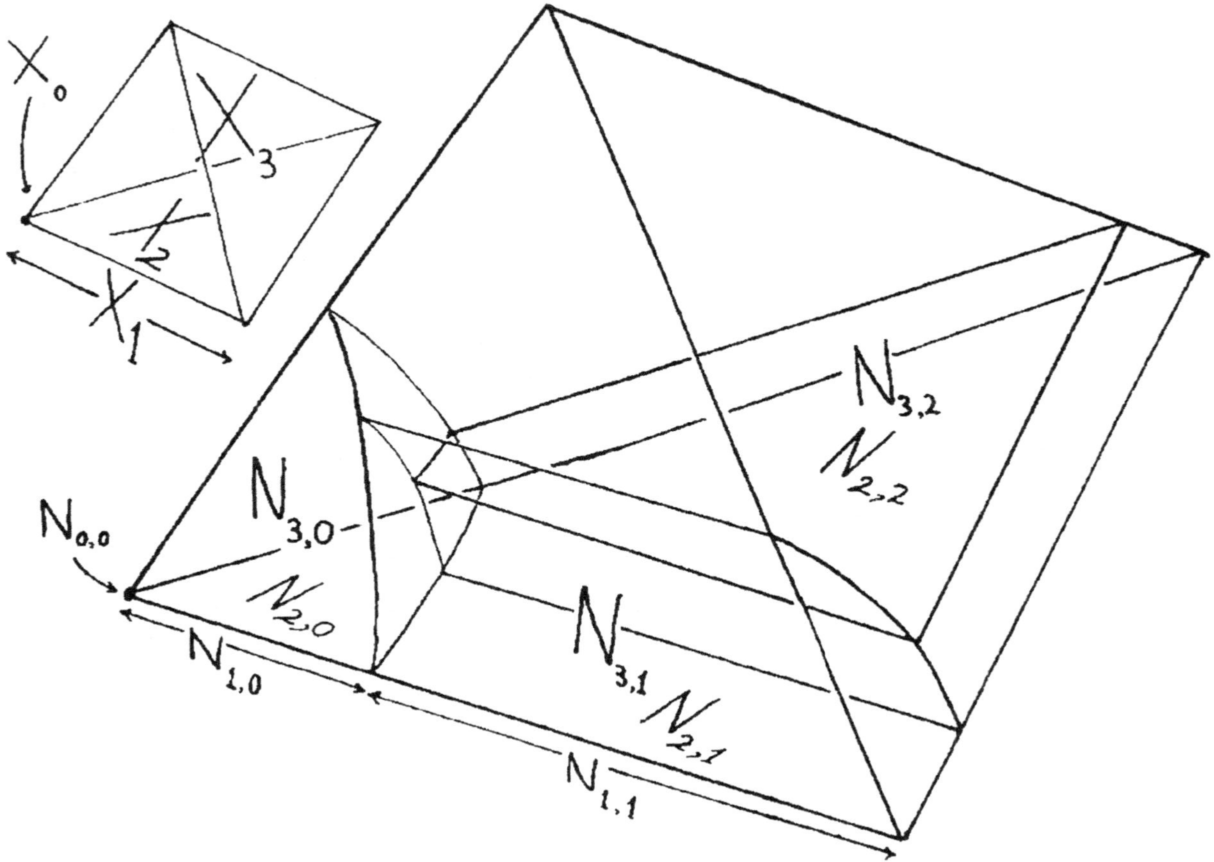

diagram 2

A normal (3,2) - ns

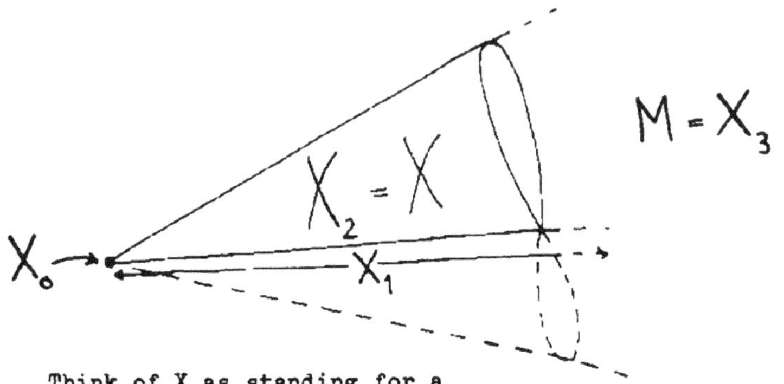

Think of X as standing for a very long cone on a figure "8". To simplify the picture, I am only drawing the upper half near the cone-point.

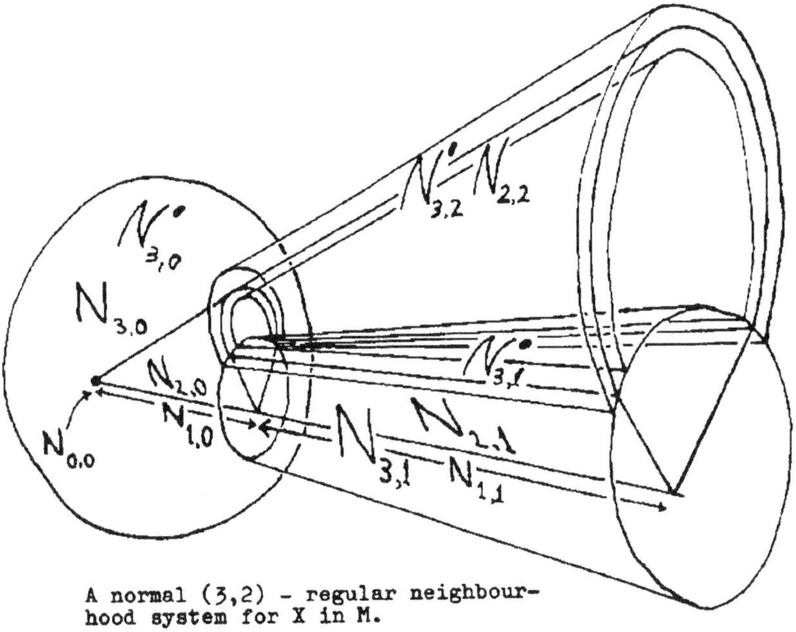

A normal (3,2) - regular neighbourhood system for X in M.

<u>diagram 3</u>

$N_{j,i} = \cup\{ s \in B' : s \in N(X_i, B'_j)$, but $s \cap X_{i-1} = \emptyset\}$,

for $0 \leq i \leq n$, $i \leq j \leq q$.

Using induction on n and Proposition 1.4 to prove the inductive step, one shows that $\underline{N} = \{N_{j,i}\}$ is the required normal (q, n)-rns.

Proof of 2.: uses induction on n and Theorems 1.1 and 1.3 to prove the inductive step.

Corollary. Given equivalent filtrations $(X_q, \ldots, X_o)$ and $(X'_r, \ldots, X'_o)$ of X, such that $X_n = X'_m$; and given $X^* <$ bdy(X) and $<$ bdy(X'). Let $\underline{N}$ be a normal(q, n)-rns for X_n in (X) with respect to X^*. Then one obtains a normal (r, m)-rns for X'_m in (X') with respect to X^* by renumbering the components of the $N_{j,i}$'s. To be precise, set

$N'_{j,i} = \cup\{ N''_{k,\ell} : N''_{k,\ell}$ is a component of $N_{k,\ell} \in \underline{N}$, $N''_{k,\ell} \cap X'_i \neq \emptyset$, $N''_{k,\ell} \cap X'_{i-1} = \emptyset\}$; and set $N'^{\bullet}_{j,i} = \cup\{N^{\bullet}_{k,\ell} \cap N''_{k,\ell}\}$.

This follows from the construction used to prove the first part of the Proposition: one simply chooses B to triangulate (X); then B also triangulates (X'). Thus one obtains a normal (q, n)-rns $\underline{M}$ and a normal (r, m)-rns $\underline{M}'$ related as stated. By part 2 of the Proposition there is an isotopy f_t of (X) carrying $\underline{M}$ to $\underline{N}$; but f_t also respects the filtration (X'), so $\underline{N}' = f_1(\underline{M}')$ is a normal (r, m)-rns related to $\underline{N}$ as stated.

Remark. The assertion of Proposition 1.11, 1 is a tautology, since the definition of a (q, n)-ns is constructive: one takes a regular neighbourhood of X_o in (X), throws it aside, takes a regular neighbourhood of what's left of X_1 in what's left of (X), and so on. The proof of the assertion is more important; it says that rather than successively taking smaller and smaller second-derived neighbourhoods, one can construct a (q, n)-ns with just one second-derived triangulation.

Let $N_{j,} = \cup\{N_{j,i} : i = 0, \ldots, \min(n, j)\}$. Then we have the filtered polyhedron $(N) = (N_{q,}, \ldots, N_{o,})$. The proof of Proposition 1.11 shows that $(N_{q,}, \ldots, N_{n+1,})$ is a regular neighbourhood of X_n in (X). Define fr N to be its frontier, filtered by $(\text{fr } N_{q,}, \ldots, \text{fr } N_{n+1,})$; note that $N_{j,} \cap \text{fr } N = \emptyset$ for $j \leq n$.

Given numbers $n \leq q$, a family of polyhedra $\underline{N} = \{N_{j,i} : 0 \leq i \leq n, \ i \leq j \leq q\}$, and a polyhedron fr N. Then $\underline{N}$ is a <u>regular</u> (q, n)-<u>neighbourhood system</u> (regular(q, n)-ns) if there is a filtered polyhedron $(X_q, \ldots, X_o)$ such that N is a normal (q, n)-ns for X_n in (X) with frontier fr N. $\underline{N}$ is a <u>weak</u> (q, n)-<u>neighbourhood system</u> (weak (q, n)-ns) if there exist a filtered polyhedron $(X_q, \ldots, X_o)$ and a normal (q, n)-ns $\underline{N}'$ for X_n in (X) such that $N_{j,i} = N'_{j,i} \cap N'_{q,o}$, for all j, i; and fr N is defined as $\text{fr } N' \cap N'_{q,o}$. $\underline{N}$ is a (q, n)-<u>neighbourhood system</u> ((q, n)-ns) if it is either a regular (q, n)-ns or a weak (q, n)-ns. In case q = n, we write: $\underline{N}$ is an n-ns.

<u>Remarks</u>

1. We could avoid introducing the auxiliary (X) by giving definitions in terms of weak regular neighbourhoods.

2. In accordance with our notion of the equivalence of filtrations, there

is no need for our indices to start from 0; what is actually used is a filtered polyhedron $(X_{q+a}, \dots, X_a)$ of length q and the subfiltration of length n. We stick to the convention that X_i and $N_{j,i}$ are $\emptyset$ for $i \lneq 0$. If $\underline{N}$ is a normal rns for X_{n+a} in (X_{q+a}), we can regard $\underline{N}$ as a (q,n)-rns, or a $(q+a, n+a)$-rns by setting $N_{j,i} = \emptyset$ for $i \lneq a$.

Given numbers $n \leq q$, a filtered polyhedron $(X_q, \dots, X_o)$, a normal (q,n)-rns $\underline{N}$ for X_n in (X), and a subpolyhedron $Z \subseteq X$. For each $N_{j,i} \in \underline{N}$, set $M_{j,i} = Z \cap N_{j,i}$, and let $\underline{M} = \{M_{j,i}\}$. We define $\underline{M}$ to be a restriction of N by induction on n:

1. $(N_{q,o}, \{M_{q,o}, N_{j,o}\})$ is a regular neighbourhood of X_o in $(X_q, \{Z, X_j\})$;
2. Set $\underline{N}' = \{N_{j,i} : i \geq 1\}$, $\underline{M}' = \{N_{j,i} : i \geq 1\}$; then $\underline{M}'$ is a restriction of $\underline{N}'$;
3. $M'_{j,} \cap N^{\bullet}_{q,o} = M^{\bullet}_{j,o}$ for $j = 1, \dots, q$.

This is stronger than simply requiring $\underline{M}$ to be a normal (q,n)-rns for Z_n in (Z). We write: $\underline{M} < \underline{N}$. Similarly we define restrictions of regular and weak (q,n)-ns's. If $Z < \mathrm{bdy}(N_{q,}, \{N_{j,i}\})$, then it follows that $\underline{M}$ is a restriction of $\underline{N}$; in this case we say: $\underline{M} < \mathrm{bdy}\, \underline{N}$.

Lemma 1.12. Given (q,n)-ns's $\underline{N}$ and $\underline{N}'$, restrictions $\underline{M} < \underline{N}$, $\underline{M}' < \underline{N}'$, and an isomorphism $f : \underline{M} \approx \underline{M}'$. Define the family $\underline{P} = \{P_{j,i} : 0 \leq i \leq n,\ i \leq j \leq q\}$ by: $P_{j,i} = N_{j,i} \cup_f N'_{j,i}$ is formed from the disjoint union by identifying $M_{j,i}$ to $M'_{j,i}$ by f, for all i, j. Then $\underline{P}$ is a (q,n)-ns.

The proof is an inductive use of Proposition 1.4.

We shall henceforth denote $\underline{P}$ by: $\underline{N} \cup_f \underline{N}'$, or if $f: \underline{M} \approx \underline{M}'$ is clear in context, by: $\underline{N} \cup \underline{N}'$. Note that $\underline{N}$, $\underline{N}'$, $\underline{M}$, $\underline{M}'$, $\underline{P}$ are either all regular or all weak (q, n)-ns's.

Corollary 1. Given a (q, n)-ns $\underline{N}$, and restrictions $\underline{M}, \underline{M}' < \underline{N}$ such that there is a restriction $\underline{P} < \underline{M}$ with $\underline{P}_- < \underline{M}'$ and $\underline{M} \cap \underline{M}' = \underline{P}$. Then $\underline{M} \cup \underline{M}' < \underline{N}$. Here $\underline{M} \cap \underline{M}' = \{M_{j,i} \cap M'_{j,i} : 0 \leq i \leq n,\ i \leq j \leq q\}$ and $\underline{M} \cup \underline{M}' = \{M_{j,i} \cup M'_{j,i}\}$.

Corollary 2. If in Corollary 1, $\underline{M}$ and $\underline{M}'$ are $<$ bdy $\underline{N}$, and $\underline{P} <$ bdy $\underline{M}$, $\underline{P}_- <$ bdy $\underline{M}'$; then $\underline{M} \cup \underline{M}' <$ bdy $\underline{N}$.

§4. Block Bundles and Blockings

There is some resemblance between a rns of a variety of a polyhedron, and a system of tubular neighbourhoods (lambeaux d'incidence) of an ensemble stratifié as defined by Thom [28]. The resemblance will be greater once we introduce bundle structures into rns's. We shall first put a bundle structure on a tubular neighbourhood of $\mathcal{X}_i$ in X, or more precisely, on $N_{i,i}$ in $N_{n,i}$. We have to allow bundles whose fibres are not spheres but quite general compact polyhedra. Even in the case that $N_{n,i}$ is a manifold and $N_{i,i}$ a submanifold, we cannot in general hope for a locally trivial fibre bundle structure - see Rourke and Sanderson [20]. Hence we have to use block bundles with arbitrary fibre. We should also like to keep track of the filtration (X) near $\mathcal{X}_i$, so we put block bundle structures on each pair $N_{i,i} \subseteq N_{j,i}$, $j = i+1, \ldots, n$. Next, we need some compatibility condition on these block bundles for each fixed i, since they are contained one in another, to make each in some sense a sub-bundle of the next. To this end, we define a "flag" of block bundles. Then for each i we have a block bundle flag to describe (roughly) a tubular neighbourhood of $\mathcal{X}_i$ in all the larger manifolds of the variety at once. Finally we need some conditions to describe how the different block bundle flags are fitted together. The axioms I offer seem reasonable, and their necessity in proof will soon be clear. Yet it is here that the mystery of polyhedra seems to me to lie; not surprisingly, since in fitting together bundles of different types we depart most from what is known about bundles.

The final definition resembles in intent and general complexity a normalized presentation of a stratified set in [28] - which is why we call our structures "stratifications".

A <u>block</u> β consists of a polyhedron $|\beta|$, called the <u>total space</u> of β, and subpolyhedra $\mathcal{G}\beta$, its <u>base</u>, $\partial\beta$, its <u>edge</u>, and $\beta^{\bullet}$, its <u>rim</u>, such that there exists a p.ℓ. isomorphism

$$h : (|\beta|, \{\mathcal{G}\beta, \partial\beta, \beta^{\bullet}\}) \to (D \times cF, \{D \times c, \text{bdy } D \times cF, D \times F\}),$$

where D is a disk and F some compact polyhedron. h is called a <u>structure</u> for β, and we shall refer to both cF and F as the <u>fibre</u> of β. (See diagram 4, p. 1.31).

An <u>isomorphism</u> between blocks β and β' is a p.ℓ. isomorphism of quadruples

$$f : (|\beta|, \{\mathcal{G}\beta, \partial\beta, \beta^{\bullet}\}) \to (|\beta'|, \{\mathcal{G}\beta', \partial\beta', \beta'^{\bullet}\}).$$

We write: $f : \beta \approx \beta'$. It follows from Lemma 1.7 that then the fibres F of β and F' of β' are p.ℓ. isomorphic.

A block γ is a <u>subblock</u> of β if:

1. $(|\gamma|, \{\mathcal{G}\gamma, \partial\gamma, \gamma^{\bullet}\} \subseteq (|\beta|, \{\mathcal{G}\beta, \partial\beta, \beta^{\bullet}\})$;
2. there are structures h for β, g for γ and an inclusion $i : G \to F$ of fibres such that this diagram commutes:

$$\begin{array}{ccc}
(|\beta|, \{\mathcal{G}\beta, \partial\beta, \beta^{\bullet}\}) & \xrightarrow[h]{} & (D \times cF, \{D \times c, \text{bdy } D \times cF, D \times F\}) \\
\cup\!| & & \Big\uparrow \text{id}(D) \times ci \\
(|\gamma|, \{\mathcal{G}\gamma, \partial\gamma, \gamma^{\bullet}\}) & \xrightarrow[g]{} & (D \times cG, \{D \times c, \text{bdy } D \times dG, D \times G\})
\end{array}$$

Note that $\mathcal{G}\gamma = \mathcal{G}\beta$.

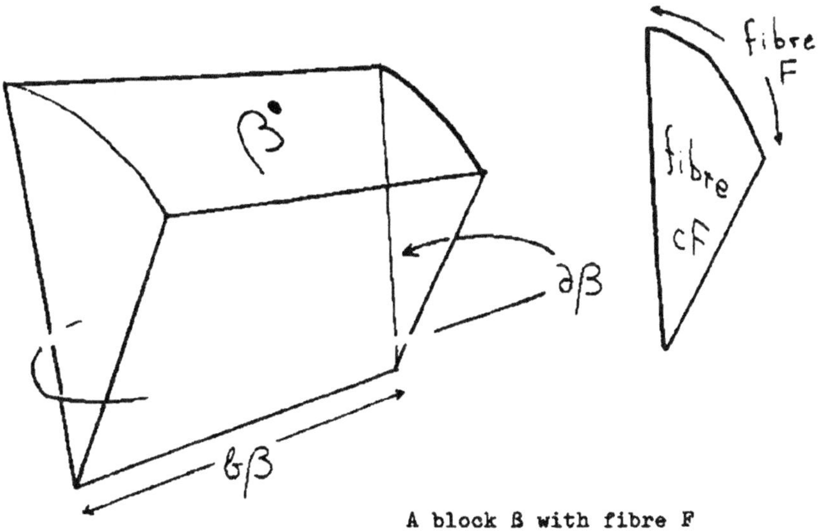

A block ß with fibre F

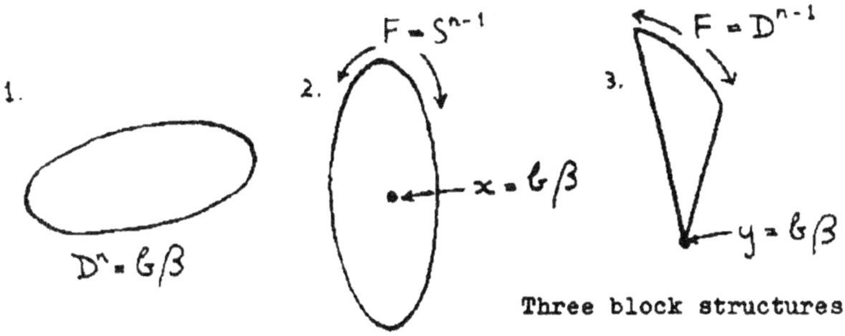

Three block structures on D^n

<u>diagram 4</u>

Remarks

1. An n-disk D^n has a canonical block structure $(D, \{D, \text{bdy } D, \emptyset\})$. With this structure, for any block β, $\mathfrak{b}\beta$ is a subblock of β.

2. Given a point $x \in \text{int } D$, D has the structure $(D, \{x, \emptyset, \text{bdy } D\})$.

3. Given a point $y \in \text{bdy } D$, D has structures $(D, \{y, \emptyset, D^{n-1}\})$, where $D^{n-1} \subseteq \text{bdy } D^n$ is a disk not containing y. (See diagram 4, p. 1.31).

4. $\partial\beta \cup \beta^{\bullet}$ is contained in the polyhedral boundary $\text{bdy}|\beta|$, but may be a proper subset. In fact $\partial\beta \cup \beta^{\bullet}$ is the complement in $\text{bdy}|\beta|$ of a regular neighbourhood in $\text{bdy}|\beta|$ of $\mathfrak{b}\beta \cap \text{bdy}|\beta|$ rel bdy $\mathfrak{b}\beta$; and $\partial\beta$ is a regular neighbourhood in $\partial\beta \cup \beta^{\bullet}$ of bdy $\mathfrak{b}\beta$. Thus for any block β, if $|\beta|$ and $\mathfrak{b}\beta$ are given, then $\partial\beta$ and $\beta^{\bullet}$ are unique up to isotopy of $|\beta|$ rel $\mathfrak{b}\beta$.

5. If X is a polyhedron and x a point of X, then $\text{st}(x, X)$ has a block structure β, whose base is canonical, namely: the smallest manifold of the intrinsic variety of $\text{st}(x, X)$. The proof will follow from Theorem 2.1. This example is in a sense the "typical" block.

Henceforth letters α, β, γ will denote blocks, and letters π, ρ, σ, τ will denote cells of a cell complex, which we shall always assume endowed with their canonical block structure, given in Remark 1. However, we shall confound σ and $|\sigma|$, and refer to both as σ.

A <u>cone block bundle</u> ξ over a cell complex K is a set of blocks such that:

cbb 1 $K = \{\mathscr{b}\beta : \beta \in \xi\}$;

cbb 2 For every $\sigma \in K$, $\mathscr{b}^{-1}\sigma = \{\beta \in \xi : \mathscr{b}\beta = \sigma\}$ has at most two elements, one of which is σ. Thus $K \subseteq \xi$. We set

$$\xi(\sigma) = \begin{cases} \mathscr{b}^{-1}(\sigma) - \sigma, & \text{if this is non-empty,} \\ \sigma, & \text{otherwise;} \end{cases}$$

cbb 3 $\{|\xi(\sigma)| - \partial\xi(\sigma) : \sigma \in K\}$ are pair-wise disjoint and cover the underlying space $|\xi| = \cup\{|\beta| : \beta \in \xi\}$ of ξ;

cbb 4 If $\rho < \sigma \in K$, then $|\xi(\rho)| \cap \xi^{\bullet}(\sigma) = \xi^{\bullet}(\rho)$;

cbb 5 $|\xi(\sigma)| \cap |\xi(\tau)| = \cup\{|\xi(\rho)| : \rho \in K, \ \rho < \sigma, \ \rho < \tau\}$ for all $\sigma, \tau \in K$.

We write: ξ/K is a cbb. The rim of ξ is $\xi^{\bullet} = \cup\{\xi^{\bullet}(\sigma) : \sigma \in K\}$.

An n-flag $(\xi) = (\xi_n, \ldots, \xi_1)$ over a cell complex K consists of cbb's $\xi_n/K, \ldots, \xi_1/K$ such that:

n-flag 1 $|\xi_n| \supseteq \ldots \supseteq |\xi_1|$ (which $\supseteq |K|$ by cbb 2 for ξ_1/K);

n-flag 2 For each $\sigma \in K$, there are compact polyhedra $F_n(\sigma) \supseteq \ldots \supseteq F_1(\sigma)$ and a p.ℓ. isomorphism $h : |\xi_n(\sigma)| \to \sigma \times cF_n(\sigma)$ which restricts to a structure $h : \xi_i(\sigma) \to \sigma \times cF_i(\sigma)$ for $i = 1, \ldots, n$.

We write: $(\xi)/K$ is an n-flag, or: $(\xi_n, \ldots, \xi_1)/K$ is an n-flag. For convenience in working with indices we shall often use the notation ξ_0 for K. We shall use $(\xi^{\bullet})$ or $(\xi)^{\bullet}$ to denote the filtered polyhedron $(\xi_n^{\bullet}, \ldots, \xi_1^{\bullet})$. Similarly $(|\xi|)$ or $|(\xi)|$ will denote the filtered polyhedron $(|\xi_n|, \ldots, |\xi_0|)$.

$(\xi)|L$, is defined as $L \cup \{\xi_i(\sigma) : \sigma \in L,\ i = 1, \ldots, n\}$. $(\xi)|L$ is clearly an n-flag over L; in particular, each $\xi_i|L$ is a cbb over L.

Given n-flags $(\xi)/K$ and $(\eta)/L$, an n-flag <u>isomorphism</u> $f : (\xi) \approx (\eta)$ is a p.ℓ. isomorphism $f : |\xi_n| \to |\eta_n|$ such that:

$f^{-1}|\eta_i| = |\xi_i|$, for $i = 0, \ldots, n$;

$f : K \to L$ is a cell complex isomorphism;

f restricts to a block isomorphism $f : \xi_i(\sigma) \to \eta_i(f\sigma)$ for all $\sigma \in K$ and all $i = 1, \ldots, n$.

If we have $L = K$, then an isomorphism $f : (\xi)/K \approx (\eta)/K$ is <u>modulo</u> K if f is the identity on K.

Given a sequence $(F) = (F_n, \ldots, F_1)$ of compact polyhedra, an n-flag $(\xi)/K$ is a <u>c(F)-block bundle</u> (abbreviated to: c(F)-bb) if for every $\sigma \in K$, $(F_n(\sigma), \ldots, F_1(\sigma)) = (F)$. (By Lemma 1.7, the sequence $(F(\sigma))$ is unique, given (ξ).)

We prove in the Appendix (Theorem I, 1.1) that given any n-flag $(\xi)/K$ and any $\sigma \in K$, then $(\xi)|\langle\sigma\rangle$ is a $c(F(\sigma))$-bb. Hence $(F(\rho)) = (F(\sigma))$ whenever $\rho < \sigma$. It follows that any n-flag $(\xi)/K$ is the disjoint union of $c(F^i)$-bb's $(\xi^i)/K^i$, where $\{K^i\}$ are the components of K. As a special case, we have the definition of a cF-bb for any compact polyhedron F, and the expression of any cbb as the disjoint union of cF^i-bb's.

<u>Remarks</u>

1. Any cell complex is a $c\emptyset$-bb over itself.

2. A q-(disk) block bundle in the sense of Rourke and Sanderson [19, I]

is a cS^{q-1}-bb. The fibre is a disk with the block structure of Remark 2, p. 1.32.

3. From the present point of view one could just as well refer to a cD^{q-1}-bb as a q-disk block bundle, where the fibre has the block structure of Remark 3, p. 1.32. This we shall henceforth do.

4. Block bundles with arbitrary fibre but without "0-section" have been defined by Casson [5]. When the fibre is required to be compact, his block bundles correspond to the rims of ours. That the two block bundle theories are then equivalent is shown by an argument similar to the proof of [19, III, Theorem 0.3 (i)].

A blocking of a (q, n)-ns $\underline{N}$ is a set of blocks $\underline{\xi}$ such that:

bs 1 $\underline{\xi}$ is the disjoint union $\cup\{(\xi_{,i}) : 0 \leq i \leq n\}$ of (q-i)-flags $(\xi_{,i}) = (\xi_{q,i}, \ldots, \xi_{i+1,i})/\xi_{i,i}$;

bs 2 $(|\xi_{,i}|) = (N_{,i})$ and $(\xi^{\bullet}_{,i}) = (N^{\bullet}_{,i})$ for $i = 0, \ldots, n$;

bs 3 If $\beta \in \xi_{q,k}$, then $\{|\beta| \cap N_{j,i} : N_{j,i} \in \underline{N}\}$ is a weak (q, n)-ns, and $(|\beta| \cap N_{n,}, \ldots, |\beta| \cap N_{k,})$ is a variety filtration of $|\beta| \cap N_{n,}$ in $(|\beta| \cap N_{q,}, \ldots, |\beta| \cap N_{n,})$;

bs 4 If $\beta \in (\xi_{,i})$ and $\gamma \in (\xi_{,k})$ with $i \leq k$, then $|\beta| \cap |\gamma| = \cup\{|\alpha| : \text{some } \alpha \in (\xi_{,k})\}$.

A set of blocks $\underline{\xi}$ is a (q, n)-block system (abbreviated to: (q, n)-bs) if there is a (q, n)-ns $|\underline{\xi}|$ of which $\underline{\xi}$ is a blocking; $|\underline{\xi}|$ is then unique. Thus we have the notation

$|\xi_{j,}| = \cup\{|\xi_{j,i}| : 0 \leq i \leq \min(j, n)\}$, and $(|\xi_{q,}|) = (|\xi_{q,}|, \ldots, |\xi_{o,}|)$.

However, I shall sometimes use $|\underline{\xi}|$ to denote $|\xi_{q,}|$.

If $\underline{\xi}$, $\underline{\eta}$ are (q, n)-bs's, then a p. ℓ. map $f : |\xi_{q,}| \to |\eta_{q,}|$ is an isomorphism of block systems if $f : |\underline{\xi}| \approx |\underline{\eta}|$ is an isomorphism of (q, n)-ns's such that for every block $\beta \in \underline{\xi}$ there is a block $f\beta \in \underline{\eta}$ satisfying: $f|\beta| = |f\beta|$ and $f : \beta \approx f\beta$ is an isomorphism of blocks.

An n-flag is in fact an (n, 0)-bs; this will sometimes simplify our notation.

We come at last to the main definitions of this paper: a stratification of a polyhedron X, and a stratification of X in Y (when $X \subseteq Y$). An (n+1, n)-stratification of X in Y consists of a disjunction (almost always a variety of X in Y; see below) $\{\Upsilon_n, \ldots, \Upsilon_o\}$ of X, a normal (n+1, n)-ns $\underline{N}$ for $\{\Upsilon\}$ in Y, and a blocking $\underline{\xi}$ of $\underline{N}$. Since $\underline{\xi}$ then determines $\underline{N}$ and $\{\Upsilon\}$, we refer to the stratification as $\underline{\xi}$, and say $\underline{\xi}$ stratifies X (or $\{\Upsilon\}$ or the associated filtration (X)) in Y. Similarly, given a weak (n+1, n)-ns $\underline{N}$ and a blocking $\underline{\xi}$ of $\underline{N}$, we say $\underline{\xi}$ is a weak (n+1, n)-stratification of $N_{n,}$ in $N_{n+1,}$.

By setting Y = X we obtain the definition of an n-stratification of X. We shall use Roman letters $\underline{J}$, $\underline{K}$, $\underline{L}$ in this case, reserving Greek letters $\underline{\xi}$, $\underline{\eta}$, $\underline{\zeta}$ primarily for the case that Y is not known to equal X. This is done to emphasize the basic theme of this paper: that every stratification $\underline{\xi}$ of X in Y contains (as a subset of blocks) a stratification $\underline{K}$ of X, and we think of $\underline{\xi}$ as a generalized bundle over $\underline{K}$ (we write $\underline{\xi}/\underline{K}$). For example, one of the main questions we shall consider is, given X, how many different pairs (Y, X) are there in which Y is a manifold regular neighbourhood of X (up to p.ℓ. isomorphism rel X)? This question can be expressed as, given a stratification $\underline{K}$ of X, how many stratifications $\underline{\xi}$ over $\underline{K}$ are there? Thus, though the definitions of, and techniques of working with, stratifications of X and stratifications of X in Y are quite similar, their significances are quite different. We shall often write the flags of $\underline{\xi}$ as $(\xi, K_{,i}) = (\xi, K_{n,i}, \ldots, K_{i+1,i})/K_{i,i}$. In this notation the symbol "ξ" is ambiguous; but if one knows to which $K_{i,i}$ a cell σ belongs, then $\xi(\sigma)$ can refer to only one block. An isomorphism modulo $\underline{K}$ between stratifications $\underline{\xi}$ and $\underline{\eta}$ over $\underline{K}$ is a bs-isomorphism $f : \underline{\xi} \approx \underline{\eta}$ which is the identity on $\underline{K}$.

Given $X \subseteq Y$ and a disjunction $\{\mathfrak{X}\}$ of X, then there is always a normal (n+1, n)-ns $\underline{N}$ for $\{\mathfrak{X}\}$ in Y, but there may well be no stratification of $\{\mathfrak{X}\}$ in Y. (For example, taking $\{\mathfrak{X}\} = \{X\}$, one has to have that X has a cbb-neighbourhood in Y, which clearly need not happen.) Our main existence theorem (Theorem 2.1) asserts that if $\{\mathfrak{X}\}$ is a variety of X in Y, then a stratification of $\{\mathfrak{X}\}$ in Y does exist. Similarly if $\underline{\xi}$ and $\underline{\xi}'$ are stratifications of a variety $\{\mathfrak{X}\}$ of X in Y, then there is an isotopy f_t of (Y, X) such that $f_1 : \underline{\xi} \approx \underline{\xi}'$, where (X) is the associated filtration to $\{\mathfrak{X}\}$. This uniqueness theorem (Theorem 3.1) is false for arbitrary disjunctions of X. (An example is described on p. 3.3.) Henceforth, therefore, unless otherwise stated, we shall assume that every stratification is of a variety of X in Y.

Having made this assumption, our p.ℓ. stratifications are closely analogous to Thom's "ensembles stratifiés" [28]. Indeed, using results of Łojasiewicz [14] one can show that an "ensemble stratifié" determines a p.ℓ. stratification which is unique in the sense just mentioned.

Any (q, n)-bs $\underline{\xi}$ is countable as a set of blocks, and $\{|\beta| : \beta \in \underline{\xi}\}$ is a locally finite family of subpolyhedra of $|\xi_{q,}|$, as described in §1. Hence there exists a standard ordering of the cells of $\underline{\xi}$; that is, a denumeration $\sigma_1, \ldots, \sigma_r, \ldots$ of the cells such that if $\rho \subseteq |\xi_{q,i}(\sigma)|$, where $\sigma \in \xi_{i,i}$, then ρ precedes σ. (Such an ordering is by no means unique.)

Let $\underline{\xi}$ be a (q, n)-bs. A subset $\underline{\eta} \subseteq \underline{\xi}$ is a sub-block system (sub-bs) if:

sub-bs 1 $\underline{\eta}$ is a (q, n)-bs with $\eta_{j,i} = \underline{\eta} \cap \xi_{j,i}$ for all i, j;

sub-bs 2 If σ is a cell of $\xi_{i,i}$ and $\sigma \in \underline{\eta}$, then $\xi_{j,i}(\sigma) \in \underline{\eta}$ for $j = i, \ldots, q$.

We write: $\underline{\eta} \lesssim \underline{\xi}$. If also:

sub-bs 3 The (q, n)-ns $|\underline{\eta}|$ is a restriction of $|\underline{\xi}|$;

then we say $\underline{\eta}$ is a restriction of $\underline{\xi}$, and write: $\underline{\eta} < \underline{\xi}$.

Examples

1. For each $k = 0, \ldots, n$, set $\mathcal{C}_k(\underline{\xi}) = \cup\{\xi_{j,i} : i > k\}$. Then $\mathcal{C}_k(\underline{\xi})$ is a (q-k, n-k)-bs by definition. For all k, $\mathcal{C}_k(\underline{\xi}) \lesssim \underline{\xi}$, but is not a restriction. $\mathcal{C}_k(\underline{\xi})$ is in some sense the "complement" of $|\xi_{k,}|$ in $|\xi|$.

2. For every cell $\sigma \in \xi_{i,i}$, set $\Lambda(\sigma, \underline{\xi}) = \{\beta \in \underline{\xi} : |\beta| \subseteq |\xi(\sigma)|\}$. It follows from bs 3 that $\Lambda(\sigma, \underline{\xi})$ is a (q, n)-bs. Hence $\Lambda(\sigma, \xi) \lesssim \underline{\xi}$; but

it is not a restriction unless $\underline{\xi}$ is a weak (q, n)-bs and $\sigma \in \xi_{o,o}$.

3. For every $\sigma \in \xi_{i,i'}$ set $\Lambda^*(\sigma, \underline{\xi}) = \{\beta \in \underline{\xi} : |\beta| \subseteq \xi^{\bullet}(\sigma)\}$. Then $\Lambda^*(\sigma, \underline{\xi}) \lesssim \Lambda(\sigma, \underline{\xi})$ and $\lesssim \underline{\xi}$, but is a restriction of neither. $\Lambda(\sigma, \underline{\xi})$ and $\Lambda^*(\sigma, \underline{\xi})$ may be regarded as a sort of "basic star" and "basic link" of σ in $\underline{\xi}$ (see p. 1.13 for the definition of a basic link in the context of polyhedra).

4. If K is a subcomplex of $\xi_{i,i'}$, then we define $\Lambda(K, \underline{\xi}) = \cup\{\Lambda(\sigma, \underline{\xi}) : \sigma \in K\}$ and $\Lambda^*(K, \underline{\xi}) = \cup\{\Lambda^*(\sigma, \underline{\xi})\}$. In particular we have $\Lambda(\partial\sigma, \underline{\xi})$, where "$\partial\sigma$" is short for "the subcomplex of $\xi_{i,i}$ that covers $\partial\sigma$". To check that $\Lambda(K, \underline{\xi})$ is a sub-bs of $\underline{\xi}$ reduces easily to checking that, if $|\Lambda(K, \underline{\xi})_{j,i'}|$ is defined to be $\cup\{|\xi(\sigma)| : \sigma \in K\} \cap |\xi_{j,i'}|$, then $|\Lambda(K, \underline{\xi})|$, defined to be $\{\Lambda(K, \underline{\xi})_{j,i'} : i' = i, \ldots, n,\ j = i', \ldots, q\}$, is a weak ns. This is done by induction on a standard ordering of K, say $\sigma_1, \ldots, \sigma_r, \ldots$ to show that $|\Lambda(\cup\{\sigma_s : s = 1, \ldots, r\}, \underline{\xi})|$ is a weak ns. The inductive step is proved by 3. above and Lemma 1.12.

5. More generally, if $\underline{\zeta} \lesssim \underline{\eta}$ and $\underline{\zeta} \lesssim \underline{\xi}$, then $\underline{\zeta} \lesssim \underline{\xi}$. Similarly if $\underline{\zeta}' < \underline{\eta}'$ and $\underline{\eta}' < \underline{\xi}'$, then $\underline{\zeta}' < \underline{\xi}'$. Note that $\underline{\zeta}$ may be a restriction of $\underline{\xi}$ even if $\underline{\zeta}$ is not a restriction of $\underline{\eta}$ nor $\underline{\eta}$ of $\underline{\xi}$. For example, take the variety filtration $(I \times I, I \times 0)$ of a square; then one can easily find 1-stratifications $\underline{J} \lesssim \underline{K} \lesssim \underline{L}$ of the illustrated subsets of the square, with $\underline{J} < \underline{L}$:

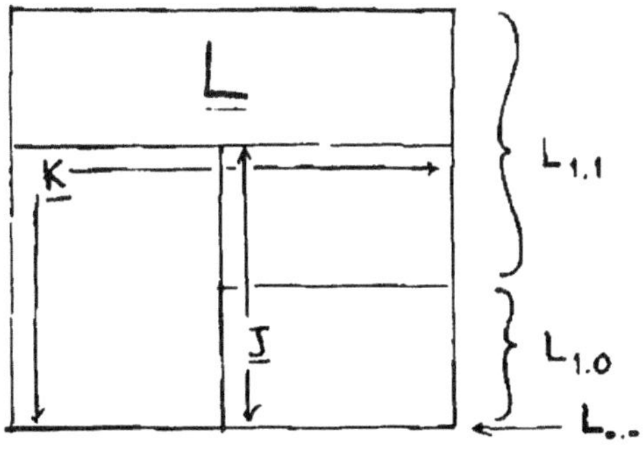

6. Given an $(n+1, n)$-stratification $\underline{\xi}/\underline{K}$ and a sub-stratification $L \precsim K$. Then there is a sub-$(n+1, n)$-stratification $\xi \restriction \underline{L} \precsim \xi$ defined by: $\underline{\xi} \restriction \underline{L} = \underline{L} \cup \{\xi(\sigma) : \sigma \in L_{i,i}$ for some $i = 0, \ldots, n\}$. If in fact $\underline{L} < \underline{K}$, then $\underline{\xi} \restriction \underline{L} < \underline{\xi}$, and we call $\underline{\xi} \restriction L$ the <u>restriction</u> of $\underline{\xi}$ to $\underline{L}$.

7. If $\underline{\eta}, \underline{\eta}' < \underline{\xi}$, then $\underline{\eta} \cup \underline{\eta}'$ and $\underline{\eta} \cap \underline{\eta}'$ are not always sub-bs's of $\underline{\xi}$; in fact, $|\underline{\eta} \cup \underline{\eta}'|$ and $|\underline{\eta} \cap \underline{\eta}'|$ are not always (q, n)-ns's. However, we have the analogue of Lemma 1.12 :

<u>Lemma 1.13.</u> Given (q, n)-bs's $\underline{\xi}$ and $\underline{\xi}'$, and restrictions $\underline{\eta} < \underline{\xi}$, $\underline{\eta}' < \underline{\xi}'$. Let $f : \underline{\eta} \approx \underline{\eta}'$ be an isomorphism of (q, n)-bs's. Let $\underline{\zeta}$ be formed from the disjoint union $\underline{\xi} \cup \underline{\xi}'$ by identifying β with $f\beta$ by f, for every $\beta \in \underline{\eta}$. Then $\underline{\zeta}$ is a (q, n)-bs, and is a blocking of the (q, n)-ns $|\underline{\xi}| \cup_f |\underline{\xi}'|$.

We shall henceforth denote $\underline{\zeta}$ by: $\underline{\xi} \cup_f \underline{\xi}'$, or if $f : \underline{\eta} \approx \underline{\eta}'$ is clear in context, by: $\underline{\xi} \cup \underline{\xi}'$. For example, if $\underline{\eta}, \underline{\eta}' < \underline{\xi}$ and there is a restriction $\underline{\zeta} < \underline{\eta}$ such that $\underline{\eta} \cap \underline{\eta}' = \underline{\zeta}$ and $\underline{\zeta} < \underline{\eta}'$; then $\underline{\eta} \cup \underline{\eta}' < \underline{\xi}$.

The notation $\underline{\xi}^* < \text{bdy}\, \underline{\xi}$ will mean:

1. $\underline{\xi}^* < \underline{\xi}$;
2. $|\underline{\xi}^*| < \text{bdy}\, |\underline{\xi}|$;
3. For every cell $\sigma \in \xi_{i,i}$, either $\Lambda^*(\sigma, \underline{\xi}) \subseteq \underline{\xi}^*$, or $\Lambda^*(\sigma, \underline{\xi}) \cap \underline{\xi}^* \subseteq \Lambda^*(\partial\sigma, \underline{\xi})$.

<u>Examples</u>

1. If $\underline{\xi}^* < \text{bdy}\, \underline{\xi}$, then for each $k = 0, \ldots, n$, $\mathcal{C}_k \underline{\xi}^* < \text{bdy}\ \mathcal{C}_k \underline{\xi}$.

2. If $\eta, \eta' <$ bdy ξ and there is a $\mathfrak{h} <$ bdy η such that $\mathfrak{h} <$ bdy η' and $\mathfrak{h} = \eta \cap \eta'$, then $\eta \cup \eta' <$ bdy ξ.

3. For each $k = 0, \ldots, n$, set $\Lambda_k^*(\xi) = \{\beta \in \xi : \beta \in \xi_{j,i}$ for some $i \underset{+}{>} k$, and $|\beta| \subseteq (\xi^\bullet_{,i'})$ for some $i' \leq k\}$. In Lemma 1.14 we shall see that Λ_k^* is a $(q-k-1, n-k-1)$-bs. For now we note that $\Lambda_o^*(\xi) = \Lambda^*(\xi_{o,o}, \xi)$ is a sub-bs of ξ, by Example 4, p. 1.39. If $|\xi|$ is a regular (q,n)-ns, then $\Lambda_o^*(\xi) <$ bdy $\mathcal{C}_o^*(\xi)$, essentially by ns 2 for $|\xi|$. Similarly $\Lambda_k^*(\mathcal{C}_{k-1}(\xi))$ is a sub-bs $<$ bdy $\mathcal{C}_k(\xi)$.

4. If $\xi^* <$ bdy ξ, then $\mathcal{C}_k(\xi^*) <$ bdy $\mathcal{C}_k(\xi)$, for $k = 0, \ldots, n$.

<u>Lemma 1.14(n)</u>. Let ξ be a (q,n)-bs with $|\xi|$ a regular ns. Then for any $k = 0, \ldots, n$: 1. $\Lambda_k^*(\xi)$ is a bs and is $<$ bdy $\mathcal{C}_k(\xi)$; and 2. $\Lambda_k^*(\xi^*) <$ bdy $\Lambda_k^*(\xi)$, if $\xi^* <$ bdy ξ.

We use induction on n. For $n = 0$, $\Lambda_k^*(\xi) = 0 = \mathcal{C}_k(\xi)$ and there is nothing to prove. Now assume Lemma 1.14(n-1):

We use decreasing induction on $j = -1, \ldots, k-1$ to prove that $\Lambda_k^*(\mathcal{C}_j(\xi)) <$ bdy $\mathcal{C}_k(\xi)$ (where $\mathcal{C}_{-1}(\xi)$ means simply ξ). For $j = k-1$, this is by Example 3 above. For the general step, $\Lambda_k^*(\mathcal{C}_j(\xi)) = \Lambda_k^*(\mathcal{C}_{j+1}(\xi)) \cup \mathcal{C}_k(\Lambda_{j+1}^*(\mathcal{C}_j(\xi)))$. Now $\Lambda_k^*(\mathcal{C}_{j+1}(\xi)) <$ bdy $\mathcal{C}_k(\xi)$ by inductive hypothesis on j; and $\mathcal{C}_k(\Lambda_{j+1}^*(\mathcal{C}_j(\xi))) <$ bdy $\mathcal{C}_k(\mathcal{C}_{j+1}(\xi))$ by Examples 3 and 4; since $\mathcal{C}_k(\mathcal{C}_{j+1}(\xi)) = \mathcal{C}_k(\xi)$, we have $\mathcal{C}_k(\Lambda_{j+1}^*(\mathcal{C}_j(\xi))) <$ bdy $\mathcal{C}_k(\xi)$. Moreover $\Lambda_k^*(\mathcal{C}_{j+1}(\xi) \cap \mathcal{C}_k(\Lambda_{j+1}^*(\mathcal{C}_j(\xi))) =$ $\Lambda_k^*(\Lambda_{j+1}^*(\mathcal{C}_j(\xi)))$; this is $<$ bdy $\Lambda_k^*(\mathcal{C}_{j+1}(\xi))$ by Lemma 1.14(n-1), 2, applied to $\Lambda_{j+1}^*(\mathcal{C}_j(\xi)) <$ bdy $\mathcal{C}_{j+1}(\xi)$, and it is $<$ bdy $\mathcal{C}_k(\Lambda_{j+1}^*(\mathcal{C}_j(\xi)))$ by

Lemma 1.14(n-1), 1. Hence, by Example 2 above, $\Lambda_k^*(\mathcal{C}_j(\underline{\xi})) < \text{bdy}\ \mathcal{C}_k(\underline{\xi})$. This is the general step in our induction on j; at the end of this induction we have proved Lemma 1.14(n), 1.

For the second part, we write $\Lambda_k^*(\underline{\xi}) = \Lambda_k^*(\mathcal{C}_o(\underline{\xi})) \cup \mathcal{C}_k(\Lambda_o^*(\underline{\xi}))$ as before, and similarly for $\underline{\xi}^*$. Then $\Lambda_k^*(\mathcal{C}_o(\underline{\xi}^*)) < \text{bdy}\ \Lambda_k^*(\mathcal{C}_o(\underline{\xi}))$ by Lemma 1.14(n-1), 2; and $\mathcal{C}_k(\Lambda_o^*(\underline{\xi})) < \text{bdy}\ \mathcal{C}_k(\Lambda_o^*(\underline{\xi}))$ by Example 4 applied twice. Similarly $\Lambda_k^*(\mathcal{C}_o(\underline{\xi}^*)) \cap \mathcal{C}_k(\Lambda_o^*(\underline{\xi}^*)) = \Lambda_k^*(\Lambda_o^*(\underline{\xi}^*)) < \text{bdy}\ \Lambda_k^*(\Lambda_o^*(\underline{\xi}))$. It follows that $\Lambda_k^*(\underline{\xi}^*) < \text{bdy}\ \Lambda_k^*(\underline{\xi})$. This proves Lemma 1.14(n), 2, and completes the proof of Lemma 1.14.

<u>Corollary.</u> Given $\underline{\xi}^* < \text{bdy}\ \underline{\xi}$ (q, n)-bs's; then for each $k = 0, \ldots, n$, $\mathcal{C}_k(\underline{\xi}^*) \cup \Lambda_k^*(\underline{\xi}) < \text{bdy}\ \mathcal{C}_k(\underline{\xi})$.

Given a (q, n)-ns $\underline{N}$ and $\underline{N}^* < \text{bdy}\ \underline{N}$, a blocking $\underline{\xi}$ of $\underline{N}$ <u>respects</u> $\underline{N}^*$ if there is a sub-bs $\underline{\xi}^* < \text{bdy}\ \underline{\xi}$ which is a blocking of $\underline{N}^*$

Chapter 2. Existence

We have already seen that given $X \subseteq Y$ and a disjunction $\{\Upsilon\}$ of X in Y, then a normal rns $\underline{N}$ for $\{\Upsilon\}$ in Y exists. We propose to show now that if $\{\Upsilon\}$ is a variety of X in Y, then a blocking of $\underline{N}$ exists; that is, that there is a stratification of $\{\Upsilon\}$ in Y. The construction is, in principle, extremely simple. It is based on Rourke and Sanderson's construction of a normal block bundle of a submanifold M in a manifold Q [19, I]. Recall that (if bdy M = bdy $Q = \emptyset$) one triangulates Q by B so that M is covered by a full subcomplex A, takes a first-derived subdivision B' of B, and takes for blocks the pairs consisting of: the dual of a simplex s of A in B', the dual of s in A'. Indeed, the definition of a stratification was motivated by the wish to describe the result of this construction in the case that M (at least) is no longer a manifold. We have already hinted at this construction in the proof of Proposition 1.11, 1, which gave a particular way of constructing normal rns's. The reader is advised to draw a few examples before plunging in to the notational difficulties that follow.

Theorem 2.1. Given polyhedra $X \subseteq Y$ and a variety $\{\Upsilon_n, \ldots, \Upsilon_o\}$ of X in Y. Then there exists a stratification $\underline{\xi}$ of $\{\Upsilon\}$ in Y, for some choice of normal $(n+1, n)$-rns $\underline{N}$ of $\{\Upsilon\}$ in Y. (It follows from Proposition 1.11, 2 that for any normal $(n+1, n)$-rns $\underline{N}'$ of $\{\Upsilon\}$ in Y there exists a blocking $\underline{\xi}'$ of $\underline{N}$.)

Addendum. If also $(Y^*, X^*) < \text{bdy}(Y, X)$ is given (the filtration of X associated to $\{\mathfrak{X}\}$ is understood), then $\underline{N}$ can be chosen to respect (Y^*, X^*), and $\underline{\xi}$ can be chosen to respect $\underline{N}^*$.

The proof will also apply to the situation in which a filtered polyhedron $(Y_q, \ldots, Y_o)$, $X \subseteq Y_o$ and a variety $\{\mathfrak{X}\}$ of X in (Y) are given.

Let $'B = {}'B_{n+1}$ be a simplicial triangulation of Y in which each X_i (of the associated filtration to $\{\mathfrak{X}\}$) is covered by a subcomplex $'B_i$; and let B be a first derived subdivision of $'B$. This is done so that various subspaces of Y will automatically be covered by full subcomplexes of B. Let B' be a first derived subdivision of B. There is a subcomplex $\text{bdy}\ 'B < {}'B$ covering $\text{bdy}(Y, X)$; we set $\text{bdy}\ 'B_j = \text{bdy}\ 'B \cap {}'B_j$. Now let

$$\xi_{j,i} = \{s^*B_j : \hat{s} \in \mathfrak{X}_i\}$$
$$\quad \cup \{t^*\text{bdy}\ B_j : \hat{t} \in \text{bdy}\ \mathfrak{X}_i\}$$

for $0 \leq i \leq n$, $i \leq j \leq q$.

Here is the block structure on s^*B_j and $t^*\text{bdy}\ B_j$: for each s (or t) there is a unique i such that $\hat{s}$ (or $\hat{t}$) $\in \mathfrak{X}_i$. If $\hat{s} \notin \text{bdy}\ \mathfrak{X}_i$, then s^*B_j has base: s^*B_i,

edge: $\cup \{r^*B_j : s \lneq r \in B_i\}$,

rim: the simplicial $\ell k(s^*B_i, s^*B_j)$.

If $\hat{s} \in \text{bdy}\ \mathfrak{X}_i$, then

s^*B_j has base: s^*B_i,

edge: $s^*\text{bdy}\ B_j \cup \{r^*B_j : s \lneq r \in B_i\}$,

rim: $\ell k(s^*B_i, s^*B_j)$.

If $\hat{t} \in \text{bdy}\ \mathfrak{I}_i$, then

$t^*\text{bdy}\ B_j$ has base: $t^*\ \text{bdy}\ B_i$,

edge: $\cup \{ r^*\text{bdy}\ B_j : t \lneq r \in \text{bdy}\ B_i \}$,

rim: $\ell k(t^*\text{bdy}\ B_i,\ t^*\text{bdy}\ B_j)$.

Then $\underline{\xi} = \cup \{ \xi_{j,i} : 0 \leq i \leq n,\ i \leq j \leq q \}$ is an $(n+1, n)$-stratification of X in (Y, X). $\underline{\xi}$ will be called dual to $'B$.

Let $s \in B_n$, and say $\hat{s} \in \mathfrak{I}_i$. Then $st(\hat{s}; B'_{n+1}, \ldots, B'_i)$ is a regular neighbourhood of $\hat{s}$ in (Y, X). Hence there are compact polyhedra $F_{n+1} \supseteq \ldots \supseteq F_{i+1}$, a disk D, and a p.ℓ. isomorphism
$p : st(\hat{s}; B'_{n+1}, \ldots, B'_i \longrightarrow D \times c\,(F_{n+1}, \ldots, F_{i+1}, \emptyset)$.
Now $st(\hat{s}; (B')) = \partial s * s^*(B'_{n+1}, \ldots, B'_1)$ (the simplicial join), and $s^*((B')) = \hat{s} * \ell k(s; (B'))$ is a cone with subcones. Applying Lemma 1.7 to $s \times s^*((B')) \underset{p.\ell}{\approx} D \times c(F, \emptyset)$, we find a disk D' and a p.ℓ. isomorphism $s^*(B'_{n+1}, \ldots, B'_i) \underset{p.\ell.}{\approx} D' \times c(F_{n+1}, \ldots, F_{i+1}, \emptyset)$. This shows that every s^*B_j has a block structure with base s^*B_i. Similarly one shows that if $\hat{t} \in \text{bdy}\ \mathfrak{I}_i$, then $t^*\text{bdy}\ B_j$ has a block structure with base $t^*\text{bdy}\ B_i$.

Now let $h : s^*B_j \to s^*B_i \times cF_j$ be a structure. Then h corresponds

$\ell k(\hat{s}, s^*B_j) \longleftrightarrow \ell k(\hat{s}, s^*B_i \times cF_j)$

$= \ell k(\hat{s}, s^*B_i) \times cF_j \cup s^*B_i \times F_j$.

On the other hand, as simplicial complexes,

$\ell k(\hat{s}, s*B_j) = st(\ell k(\hat{s}, s*B_i), \ \ell k(\hat{s}, s*B_j))$

$\qquad\qquad \cup \ \ell k(s*B_i, s*B_j).$

Pseudo-radial projection from $\hat{s}$ gives a simplicial isomorphism

$f : \ell k(\hat{s}, s*B_j) \to [\ell k(s, B_j)]'$ (the first derived), under which

$\ell k(\hat{s}, s*B_i) \longleftrightarrow [\ell k(s, B_i)]'$;

$st(\ell k(\hat{s}, s*B_i), \ \ell k(\hat{s}, s*B_j)) \longleftrightarrow$

$\qquad st(\ell k(s, B_i), \ [\ell k(s, B_j)]');$

$\ell k(s*B_i, s*B_j) \longleftrightarrow \quad (\ell k(s, B_i), \ [\ell k(s, B_j)]').$

(See diagram 5, p. 2.5)

By definition of a regular neighbourhood, and by virtue of the isomorphism f, $st(\ell k(\hat{s}, s*B_i), \ \ell k(\hat{s}, s*B_j))$ is a regular neighbourhood of $\ell k(\hat{s}, s*B_i)$ in $\ell k(\hat{s}, s*B_j)$. By uniqueness of regular neighbourhoods, we may assume that h corresponds

$st(\ell k(\hat{s}, s*B_i), \ \ell k(\hat{s}, s*B_j)) \longleftrightarrow \ell k(\hat{s}, s*B_i) \times cF_j,$

and hence $\ell k(s*B_i, s*B_j) \longleftrightarrow s*B_i \times F_j$. Now if $\hat{s} \in \text{int } X_i$, which is a manifold without boundary, then $\ell k(\hat{s}, s*B_i)$ is a p.ℓ sphere and equals bdy $s*B_i$. Hence $\cup\{r*B_i : s \underset{+}{<} r \in B_i\}$ is a cell complex structure on bdy $s*B_i$; and under h,

$\qquad \cup\{r*B_j : s \underset{+}{<} r \in B_i\} \longleftrightarrow \partial s*B_j,$

$\qquad \ell k(s*B_i, s*B_j) \quad \longleftrightarrow (s*B_j)^{\bullet}$.

This checks the block structure on $s*B_j$ if $\hat{s} \in \text{int } X_i$.

A similar calculation checks the block structure on t*bdy B_j, when $\hat{t} \in \text{bdy } X_i$. However, if $\hat{s} \in \text{bdy } X_i$, we have the

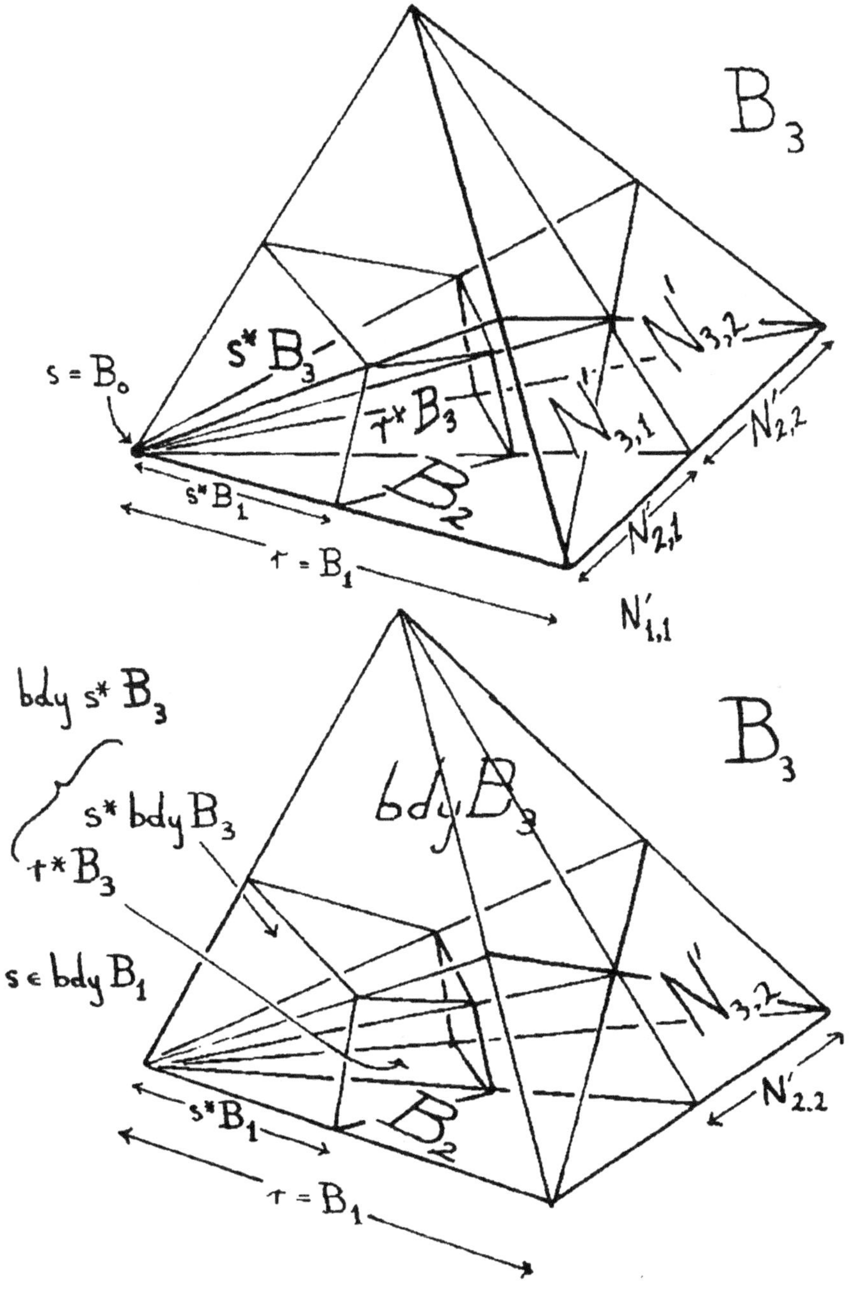
B3
s = B0
s*B3
t*B3
N'3,2
N'3,1
N'2,2
N'2,1
N'1,1
B2
s*B1
t = B1
B3
bdy s* B3
bdy B3
s* bdy B3
t*B3
s ∈ bdy B1
N'3,2
N'2.2
B2
s*B1
t = B1

diagram 5

difficulty that $\ell k(\hat{s}, s*B_i)$ is a p.ℓ. disk, not a sphere, and $\text{bdy } s*B_i = \ell k(\hat{s}, s*B_i) \cup s*\text{bdy } B_i$. We make a further calculation; there is a block structure $h : s*\text{bdy } B_j \to s*\text{bdy } B_i \times cF_j$. There is also a p.$\ell$. isomorphism

$f : s*(B_j, B_i) \to s*\text{bdy}(B_j, B_i) \times I.$

Combining with h gives

$h'' : s*B_j \longrightarrow s*\text{bdy } B_i \times I \times cF_j$, extending

$h : s*\text{bdy } B_j \longrightarrow s*\text{bdy } B_i \times 0 \times cF_j$. Now h corresponds

$$\begin{aligned} \ell k(\hat{s}, s*B_j) \longleftrightarrow (\partial\, s*\text{bdy } B_i \times I \\ \cup\ s*\text{bdy } B_i \times 1) \times cF_j \\ \cup\ s*\text{bdy } B_i \times I \times F_j. \end{aligned}$$

The previous argument, using pseudo-radial projection and the uniqueness of regular neighbourhoods, shows that we may choose h'' still to extend h and also such that

$$\begin{aligned} h : \text{st}(\ell k(\hat{s}, s*B_i), \ell k(\hat{s}, s*B_j)) \longleftrightarrow \\ (\partial\, s*\text{bdy } B_i \times I \cup s*\text{bdy } B_i \times 1) \times cF_j \end{aligned}$$

and $\ell k(s*B_i, s*B_j) \longleftrightarrow s*\text{bdy } B_i \times I \times F_j$.

Hence $s*B_j$ has edge: $s*\text{bdy } B_j \cup \text{st}(\ell k(\hat{s}, s*B_i), \ell k(\hat{s}, s*B_j))$ and rim: $\ell k(s*B_i, s*B_j)$. From this follows the desired block structure on $s*B_j$.

Let A, C be any of the B_j or $\text{bdy } B_j$. Then $r*A \cap s*C = \langle r, s\rangle*(A \cap C)$, for any $r, s \in B$; note that $A \cap C$ is again one of the B_j or $\text{bdy } B_j$. Also if $\hat{r} \in \Sigma_i$, $\hat{s} \in \Sigma_k$ and

$k \geq i$, then $\langle r, s\rangle^\wedge \in \mathcal{I}_k$.

This observation makes it easy to verify axioms about the intersection of blocks. We leave the reader to check that each $\xi_{j,i}$ is a cbb, and each $(\xi_{,i})$ a flag. The proof of Proposition 1.11, 1 shows that $\underline{N} = \{|\xi_{j,i}|\}$ is a normal $(n+1, n)$-rns for X in (Y, X). The axioms for a blocking are easily checked, except perhaps part of bs 3: that for every simplex $s \in B_n$, $\underline{N}'' = \{(s*B)^\bullet \cap N_{j,i}\}$ is a normal $(n+1, n)$-rns for $(s*B_n)^\bullet$ in $(s*(B))^\bullet$. The proof uses pseudo-radial projection from $\hat{s}$ again to give a simplicial isomorphism $f : \ell k(s*B_i, s*B) \to [\ell k(s, B)]'$. Then $\ell k(s, B_n)$ has the normal $(n+1, n)$-rns $\underline{N}'$ in $\ell k(s, (B))$ dual to the simplicial triangulation $\ell k(s, B)$. $\underline{N}'$ was constructed in the proof of Proposition 1.11, 1; $N'_{j,i} = \cup\{r \in [\ell k(s, B)]'$:
$r \in N(\ell k(s, B_i), [\ell k(s, B_j)]')$, but
$r \cap |\ell k(s, B_{i-1})| = \emptyset\}$. But f gives an isomorphism of ns's, $f : \underline{N}'' \approx \underline{N}'$. So $\underline{N}''$ is a normal $(n+1, n)$-rns, as required.

This completes the proof of Theorem 2.1.

Proof of Addendum: We are now given $(Y^*, X^*) < \mathrm{bdy}(Y, X)$. Set $Y^{**} = \mathrm{cl}[\mathrm{bdy}(Y, X) - Y^*]$, and let (Y^{**}, X^{**}) be the restriction filtration. Let $'B$ be a simplicial triangulation of Y in which Y^* and each X_i is covered by a full subcomplex, $'B^*$ and $'B_i$, of $'B$. Since, by Theorem 1.6, 2, $\mathrm{bdy}(Y, X)$ is covered by a subcomplex of $'B$, it follows that Y^{**} is covered by a subcomplex $B^{**} < B$. Take a collar neighbourhood $Y^* \times I$ of (Y^*, X^*) in (Y, X). Then

$\mathrm{bdy}(Y, X) \cap \mathrm{bdy}(Y^* \times I, X^* \times I) = Y^* \times 0 \cup \mathrm{bdy}(Y^*, X^*) \times I$.

Hence $Y \cap Y'' = \mathrm{bdy}(Y^*, X^*)$, and is thus covered by the subcomplex $\mathrm{bdy}\, B^* = B^* \cap B''$ of B. Similarly,

$\mathrm{bdy}^o(Y, X) \cap \mathrm{bdy}^o(Y^* \times I, X^* \times I) = Y^* \times 0 \cup \mathrm{bdy}^o(Y^*, X^*) \times I$.

Hence, though $\mathrm{bdy}(Y^*, X^*)$ is not necessarily collared in (Y'', X'') (for example, take $X = Z \times I$, $X^* = Z \times 0$, where Z is the region of the plane bounded by a figure "8"), we do have that $\mathrm{bdy}^o(Y^*, X^*)$ is "locally collared" in (Y'', X'') in the sense that: if $s \in B_n$ with $\hat{s} \in \mathrm{bdy}_{Y^*} X^*$, then $\mathrm{st}(\hat{s}; \mathrm{bdy}(Y^*, X^*))$ is collared in $\mathrm{st}(\hat{s}; Y'', X''))$. By definition of bdy^o, it is also collared in $\mathrm{st}(\hat{s}; Y^*, X^*))$. And

$\mathrm{st}(\hat{s}; (Y'', X'')) \cup \mathrm{st}(\hat{s}; (Y^*, X^*)) = \mathrm{st}(\hat{s}; \mathrm{bdy}(Y, X))$

is collared in $\mathrm{st}(\hat{s}; (Y, X))$, since $\hat{s} \in X^* \subseteq \mathrm{bdy}_Y X$. Using these collars, the methods of proving Theorem 2.1 apply without further innovation to prove the Addendum. The stratification $\underline{\eta}$ that we obtain is defined by:

$$\eta_{j,i} = \{s*B_j : \hat{s} \in \mathfrak{X}_i\} \cup \{s*B_j^* : \hat{s} \in \mathfrak{X}_i^*\}$$
$$\cup \{s*B_j'' : \hat{s} \in \mathrm{bdy}_Y^o X \cap \mathfrak{X}_i''\} \cup \{s*\mathrm{bdy}\, B_j^* : \hat{s} \in \mathrm{bdy}\, \mathfrak{X}_i^*\} .$$

The block structures on these blocks are:

$s*B_j$ has base: $s*B_i$;

and edge: $\cup\{r*B_j : s \underset{+}{<} r \in B_i\}$

$\cup\{s*B_j^*,\ s*B_j'',\ s*\mathrm{bdy}\, B_j^*\}$.

$s*B_j^*$ has base: $s*B_i^*$;

and edge: $\cup\{r*B_j^* : s \underset{+}{<} r \in B_i^*\}$

$\cup\ s*\mathrm{bdy}\, B_j^*$.

$s*B_j^{\cdot\cdot}$ has base: $s*B_i^{\cdot\cdot}$;

and edge: $\cup\{r*B_j^{\cdot\cdot} : s \underset{+}{<} r \in B_i^{\cdot\cdot}\}$

$\cup\, s*\mathrm{bdy}\, B_j^*$.

$s*\mathrm{bdy}\, B_j^*$ has base: $s*\mathrm{bdy}\, B_i^*$;

and edge: $\cup\{r*\mathrm{bdy}\, B_j^* : s \underset{+}{<} r \in \mathrm{bdy}\, B_i^*\}$.

In all cases, $s*A$ has rim: the simplicial $\ell k(\mathscr{b}s*A, s*A)$. $\underline{\eta}$ will be called dual to 'B with respect to (Y^*, X^*); or, if X^* is given $< \mathrm{bdy}_Y X$, and Y^* is then chosen so that $(Y^*, X^*) < \mathrm{bdy}(Y, X)$, then we say: $\underline{\eta}$ is dual to 'B with respect to X^*.

Let $(\xi)/K$ and $(\eta)/L$ be n-flags, and $J < K$ a subcomplex. Then (η) is a subdivision of (ξ) rel J (or rel$(\xi) \upharpoonright J)$ if:

subdiv 1 $|L| = |K|$ and $|\eta_i| = |\xi_i|$, for $i = 1, \ldots, n$;

subdiv 2 for each cell $\sigma \in K$,

$|(\xi)(\sigma)| = \cup\{|(\eta)(\tau)| : \tau \in L, \tau \subseteq \sigma\}$, and

$(\xi^\bullet)(\sigma) = \cup\{(\eta^\bullet)(\tau)\}$;

subdiv 3 $J < L$.

Now let $\underline{\xi}$ and $\underline{\eta}$ be (q, n)-bs's and $\underline{\xi}' \lesssim \underline{\xi}$. Then $\underline{\eta}$ is a subdivision of $\underline{\xi}$ rel $\underline{\xi}'$ if:

for each $i = 0, \ldots, n$, the flag $(\eta_{,i})/\eta_{i,i}$ is a subdivision of $(\xi_{,i})/\xi_{i,i}$ rel $\xi'_{i,i}$.

More generally, if $\underline{\xi}$ and $\underline{\eta}$ are (q, n)-bs's and $\underline{\xi}' \lesssim \underline{\xi}$, a p.$\ell$. isomorphism $f : |\underline{\eta}| \to |\underline{\xi}|$ will be sometimes called a subdivision of $\underline{\xi}$ rel $\underline{\xi}'$ if $f(\underline{\eta})$ is a subdivision of $\underline{\xi}$ rel $\underline{\xi}'$. (See diagram 6, p. 2.10).

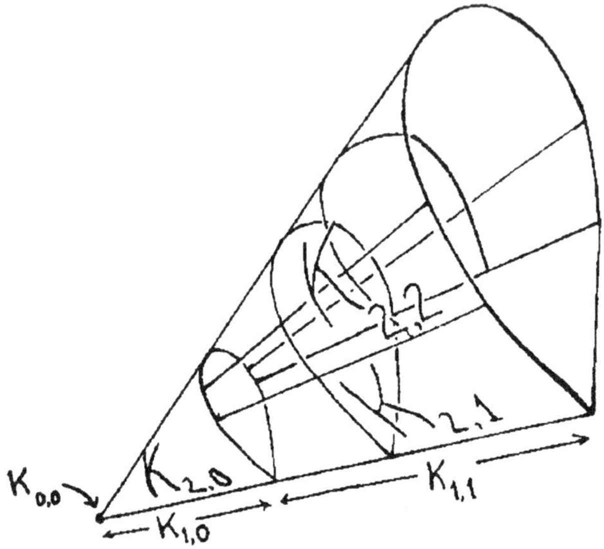

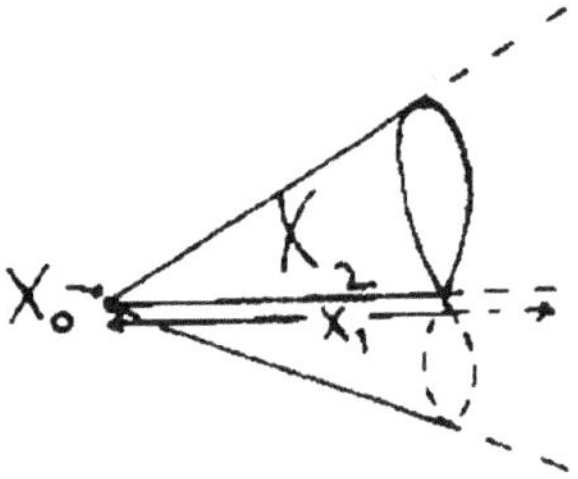

A 2-stratification $\underline{K}$ on $X = X_2$;
see diagram 3, p. 1.24

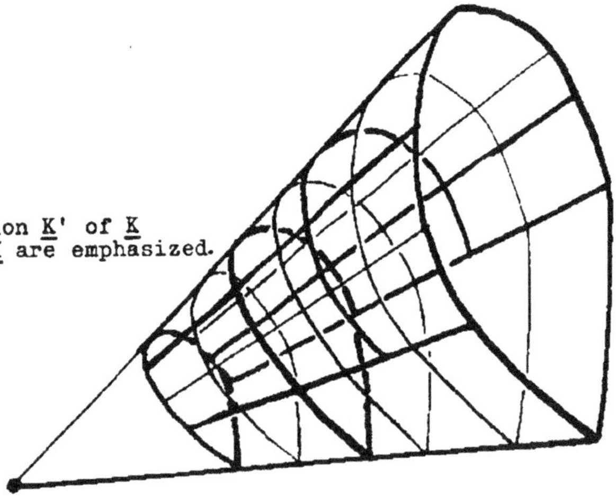

A subdivision $\underline{K}'$ of $\underline{K}$
Blocks of $\underline{K}$ are emphasized.

<u>diagram 6</u>

Let $\underline{\eta}$ be a subdivision of $\underline{\xi}$, and let $\underline{\xi}' \lesssim \underline{\xi}$. Set $\underline{\eta}\langle\underline{\xi}'\rangle = \{\beta \in \underline{\eta} : |\beta| \subseteq |\underline{\xi}'|\}$. Then $\underline{\eta}\langle\underline{\xi}'\rangle \lesssim \underline{\eta}$ and is a subdivision of $\underline{\xi}'$. Further, if $\underline{\xi}' < \operatorname{bdy} \underline{\xi}$, then $\underline{\eta}\langle\underline{\xi}'\rangle < \operatorname{bdy} \underline{\eta}$. In particular, if $\underline{\eta}$ is a subdivision of $\underline{\xi}$ rel $\underline{\xi}'$, then $\underline{\xi}' \lesssim \underline{\eta}$.

If $\underline{\eta}$ is a subdivision of $\underline{\xi}$ rel $\underline{\xi}'$, and $\underline{\zeta}$ is a subdivision of $\underline{\eta}$ rel $\underline{\xi}'$ (we have just seen that $\underline{\xi}' \lesssim \underline{\eta}$), then $\underline{\zeta}$ is a subdivision of $\underline{\xi}$ rel $\underline{\xi}'$.

Let $\underline{\xi}$ be a (q, n)-bs, $\underline{\xi}' \lesssim \underline{\xi}$, and $\underline{\eta}'$ a subdivision of $\underline{\xi}'$. Then there is a (q, n)-bs $\underline{\eta}$ defined as the set of blocks $(\underline{\xi} - \underline{\xi}') \cup \underline{\eta}'$; and $\underline{\eta}$ is a subdivision of $\underline{\xi}$. $\underline{\eta}$ will be denoted henceforth by: $\underline{\xi} + \underline{\eta}'$.

To give an example of a subdivision, take $X \subseteq Y$ and a variety filtration $(X_n, \ldots, X_o)$ of X in Y, and let $(Y^*, X^*) < \operatorname{bdy}(Y, X)$. Let $'B$ be a simplicial triangulation of Y in which Y^* and each X_i are covered by subcomplexes. We can construct the $(n+1, n)$-stratifications $\underline{\eta}/\underline{L}$ of X in Y dual to $'B$ with respect to X^*, and $\underline{\xi}/\underline{K}$ dual to $'B$. Then $\underline{\eta}$ is a subdivision of $\underline{\xi}$.

<u>Lemma 2.2.</u> Let $(\xi) = (\xi_{n+1}, \xi_n, \ldots, \xi_1)/K$ be an $(n+1)$-flag such that for every $\sigma \in K$, if $F_{n+1}(\sigma) \supseteq \ldots \supseteq F_1(\sigma)$ are the fibres of $\xi_{n+1}(\sigma), \ldots, \xi_1(\sigma)$, then $(F_n(\sigma))$ is a variety filtration of $F_n(\sigma)$ in $F_{n+1}(\sigma)$. Then there exists a weak $(n+1, n)$-bs $\underline{\eta}/\underline{L}$ such that the flag $(\eta, L_{,o})/L_{o,o} = (\xi)/K$.

$\underline{\eta}$ will be called an $(n+1, n)$-bs <u>completion</u> of (ξ).

The proof of the lemma also enables us to find a (q, n)-bs completion (which we leave the reader to define) of a q-flag $(\xi_q, \ldots, \xi_1)/K$ such that for every $\sigma \in K$, $(F_n(\sigma), \ldots, F_1(\sigma))$ is a variety filtration of $F_n(\sigma)$ in $(F_q(\sigma), \ldots, F_n(\sigma))$.

Let 'B be a simplicial triangulation of $(\xi^\bullet)$ such that $\xi_i^\bullet(\sigma)$ is covered by a subcomplex $'B_i(\sigma) < {}'B$ for every $\sigma \in K$ and each $i = 1, \ldots, n+1$. For every $\sigma \in K$, let $W(\sigma)$ be the filtered polyhedron $(\xi^\bullet) \restriction \partial\sigma$, that is, $\cup\{(\xi^\bullet(\rho)) : \rho \in K, \rho \lneq \sigma\}$. Let $V^o(\sigma)$ be the filtered polyhedron $\text{bdy}^o(\xi^\bullet(\sigma)) - (W(\sigma))$ and let $V(\sigma) = \text{cl}[V^o(\sigma)]$. As in the proof of the Addendum to Theorem 2.1, one shows that $V(\sigma)$ is covered by a subcomplex $'A(\sigma) < {}'B(\sigma)$.

Let B be a first derived subdivision of 'B. Define $\underline{\eta}$ to be the set of blocks $(\xi) \cup \underline{\eta}^\wedge$, where

$$\begin{aligned}\underline{\eta}^\wedge = \{ & s*B_j(\sigma) : s \in B_j(\sigma) \text{ for some } \sigma \in K \text{ and} \\ & \text{some } j = 1, \ldots, n\} \\ \cup\, \{ & t*A_j(\sigma) : t \in B_j(\sigma) \text{ and } \hat{t} \in V^o(\sigma)\}.\end{aligned}$$

The structures on the blocks of $\underline{\eta}^\wedge$ are: if $\hat{s} \in \xi_i^\bullet(\sigma) - \xi_{i-1}^\bullet(\sigma)$ for some i, then

$s*B_j(\sigma)$ has base: $|s*B_i(\sigma)$;

edge: $\cup\{r*B_j(\sigma) : s \lneq r \in B_i(\sigma)\}$

$\cup\{s*B_j(\rho) : \rho \lneq \sigma\}$;

rim: $\ell k(s*B_i(\sigma), s*B_j(\sigma))$.

If $\hat{t} \in V_i^o(\sigma) - V_{i-1}^o(\sigma)$, then

$t*A_j(\sigma)$ has base: $t*A_i(\sigma)$;

edge: $\cup\{r*A_j(\sigma) : t \underset{+}{<} r \in A_i(\sigma)\}$

$\cup\{t*A_j(\rho) : \rho \underset{+}{<} \sigma\}$;

rim: $\ell k(t*A_i(\sigma), t*A_j(\sigma))$.

The proof that $\underline{\eta}$ is an $(n+1, n)$-bs uses induction on the dimension of K. For every $\sigma \in K$, let $\sigma\underline{\eta}^{\wedge} = \{s*B_j(\sigma), t*A_j(\sigma)$ which are contained in $(\xi^{\bullet}(\sigma))\}$. We first show that $\sigma\underline{\eta}^{\wedge}$ is an $(n, n-1)$-stratification. Note that $W(\sigma) < \text{bdy}(\xi^{\bullet}(\sigma))$ and $W(\sigma)$ is covered by the subcomplex $\cup\{'B(\rho) : \rho \underset{+}{<} \sigma\}$ of $'B(\sigma)$. Let $\sigma\underline{\xi}$ be an $(n, n-1)$-stratification of $(\xi^{\bullet}_n(\sigma))$ in $\xi^{\bullet}_{n+1}(\sigma)$ which is dual to $'B(\sigma)$ with respect to $(W(\sigma))$; thus $(W_n(\sigma))$ has an $(n, n-1)$-stratification $\sigma\underline{\xi}^* < \text{bdy}\,\sigma\underline{\xi}$ in $W(\sigma)$. On the other hand, applying inductive hypothesis to the flag $(\xi) \vdash \langle\partial\sigma\rangle$, we have the $(n, n-1)$-stratification $\partial\sigma\underline{\eta}^{\wedge} = \cup\{\rho\underline{\eta}^{\wedge} : \rho \underset{+}{<} \sigma\}$ of $(W_n(\sigma))$ in $W(\sigma)$. From our laborious descriptions of $\partial\sigma\underline{\eta}^{\wedge}$ and $\sigma\underline{\xi}^*$ it follows readily that $\partial\sigma\underline{\eta}^{\wedge}$ is a subdivision of $\sigma\underline{\xi}^*$, and that $\sigma\underline{\eta}^{\wedge} = \sigma\underline{\xi}^* + \partial\sigma\underline{\eta}^{\wedge}$. Hence $\sigma\underline{\eta}^{\wedge}$ is an $(n, n-1)$-stratification of $(\xi^{\bullet}_n(\sigma))$ in $\xi^{\bullet}_{n+1}(\sigma)$, and $\partial\sigma\underline{\eta}^{\wedge} < \text{bdy}\,\sigma\underline{\eta}^{\wedge}$. Now take a standard ordering $\sigma_1, \ldots, \sigma_m, \ldots$ of the cells of K. By induction on m one shows that $\cup\{\sigma_k\underline{\eta}^{\wedge} : k = 1, \ldots, m\}$ is an $(n, n-1)$-stratification of $\cup\{(\xi^{\bullet}_n(\sigma_k))\}$ in $\cup\{\xi^{\bullet}_{n+1}(\sigma_k)\}$; the inductive step is proved using Lemma 1.13. Thus $\underline{\eta}^{\wedge}$ is an $(n, n-1)$-stratification of $(\xi^{\bullet}_n)$ in $\xi^{\bullet}_{n+1}$. It is now straight-forward to check that $\underline{\eta} = (\xi) \cup \underline{\eta}^{\wedge}$ is an $(n+1, n)$-bs, as required.

Chapter 3. Uniqueness

Uniqueness theorems are more difficult to prove than existence theorems, even in the case of normal disk block bundles of submanifolds in manifolds (see Rourke and Sanderson [19, I]). First we must decide, unique in what sense? That is the purpose of this preliminary discussion.

Recall Rourke and Sanderson's uniqueness theorem: Let $V \subseteq M$ be a submanifold of a manifold, and let ξ/K, η/L be normal block bundles of V in M. Take a common cell-complex subdivision J of K and L, and let ξ'/J, η'/J be subdivisions of ξ and η. Then there is an isotopy f_t of M rel V such that $f_1 : \xi' \approx \eta'$ is a block bundle isomorphism. Thus in general, given a subpolyhedron $X \subset M$, a variety filtration (X) of X in M, and stratifications $\underline{\xi}/\underline{K}$, $\underline{\eta}/\underline{L}$ for (X) in M, we must be prepared to subdivide $\underline{\xi}$ and $\underline{\eta}$.

Consider the normal block bundles $\xi_{n+1,0}/K_{0,0}$ and $\eta_{n+1,0}/L_{0,0}$ of X_0 in M (recall that X_0 is a submanifold of M). Again let $\xi'_{n+1,0}/J$ and $\eta'_{n+1,0}/J$ be subdivisions of $\xi_{n+1,0}$ and $\eta_{n+1,0}$. The question arises: can $\xi'_{n+1,0}$ and $\eta'_{n+1,0}$ be chosen to come from subdivisions $\underline{\xi}'$ of $\underline{\xi}$ and $\underline{\eta}'$ of $\underline{\eta}$? One can readily find a flag subdivision $(\xi'_{n+1,0}, \ldots, \xi'_{1,0})/J$ of $(\xi_{n+1,0}, \ldots, \xi_{1,0})/K_{0,0}$ subordinate to $\xi'_{n+1,0}$; but the problem is that the new rims $\{\alpha'^{\bullet} : \alpha' \in \xi'_{n+1,0}\}$ and the old blocks $\{\beta : \beta \in \underline{\xi}$ with $|\beta| \subseteq \xi^{\bullet}_{n+1,0}\}$ need not fit together well.

Assume, however, that we have found subdivisions $\underline{\xi}'/\underline{K}'$ of $\underline{\xi}$

and $\underline{\eta}'/\underline{L}'$ of $\underline{\eta}$ such that $K'_{o,o} = J = L'_{o,o}$. Then Rourke and Sanderson's proof can be modified to show that there is an isotopy f_t of M rel X_o such that $f_1 : (\xi'_{n+1,0}, \ldots, \xi'_{1,0})/J \approx (\eta'_{n+1,0}, \ldots, \eta'_{1,0})/J$ is a flag isomorphism. Note that if the block bundles $\xi'_{n,o}/J$ and $\eta'_{n,o}/J$ are not equal, we cannot keep X fixed by f_t (even if the normal rns's $|\underline{\xi}|$ and $|\underline{\eta}|$ are equal). So our uniqueness theorem for stratifications must allow subdivisions and isotopies of M that shift (X) within itself. These two relations are in fact sufficient. The problem that I mentioned before recurs, but can again be solved.

We shall establish three uniqueness theorems by means of Theorem 3.1:

1. Let (X) be a variety filtration of X, and given stratifications $\underline{K}$, $\underline{L}$ of (X). Then there are subdivisions $\underline{K}'$ of $\underline{K}$, $\underline{L}'$ of $\underline{L}$ and an isotopy f_t of (X) such that $f_1 : \underline{K}' \approx \underline{L}'$.

2. Let (X) be a variety filtration of X in Y, and given stratifications $\underline{\xi}$, $\underline{\eta}$ of (X) in Y. Then there are subdivisions $\underline{\xi}'$ of $\underline{\xi}$, $\underline{\eta}'$ of $\underline{\eta}$ and an isotopy f_t of (Y, X) such that $f_1 : \underline{\xi}' \approx \underline{\eta}'$.

3. Let (X) be a variety filtration of X in Y, and given stratifications $\underline{\xi}/\underline{K}$, $\underline{\eta}/\underline{K}$ of (X) in Y that agree on (X). Then there is an isotopy f_t of Y rel X such that $f_1 : \underline{\xi} \approx \underline{\eta}$.

Our technique relies on Rourke and Sanderson's theorem, though we do not explicitly need its proof. The proof definitely requires the base space of the given block bundles to be a manifold. And in fact our uniqueness theorem fails if (X) is allowed to be any filtration of X in Y. We outline an example: According to Rourke and Sanderson [21] one can find, whenever ℓ is odd and q, $r \geq 3$, normal cS^{r-1}-block bundles ξ/K, η/H for $M^{\ell+q} = S^{\ell} \times D^q \times 0$ in $Q^{\ell+q+r} = S^{\ell} \times D^q \times D^r$ such that: $L^{\ell} = S^{\ell} \times 0 \times 0$ is covered by a subcomplex K' of both K and H, and $\xi \restriction K' = \eta \restriction K'$ is a normal cS^{r-1}-block bundle for L in $N = S^{\ell} \times 0 \times D^r$; but there exists <u>no</u> normal cS^{r-1}-block bundle μ/P for $M \times I$ in $Q \times I$ which extends subdivisions of $\xi \times 0$ and $\eta \times 1$ and which has a restriction $\mu \restriction P'$ with $|P'| = L \times I$ and $\mu \restriction P'$ a normal cS^{r-1}-block bundle for $L \times I$ in $N \times I$.

Now let Y be obtained from the disjoint union $Q \cup N \times I$ by identifying N with $N \times 0$; and let $X \subseteq Y$ be the image of $M \cup L \times I$. Then $\xi^{\wedge}/K^{\wedge} = \xi \cup (\xi \restriction K') \times I$ and $\eta^{\wedge}/H^{\wedge} = \eta \cup (\eta \restriction K') \times I$ are both (1,0)-block systems for X in Y. Suppose subdivisions $\xi^{\vee}/K^{\vee}$ of $\xi^{\wedge}$, $\eta^{\vee}/H^{\vee}$ of η and an isotopy f_t of (Y, X) are given such that $f_1 : \xi^{\vee} \approx \eta^{\vee}$. Regarding f_t as a p.ℓ. isomorphism $f_t : Y \times I \to Y \times I$, the block bundle $f_t(\xi^{\vee} \times I/K^{\vee} \times I)$ forms a concordance between $(\xi^{\vee}/K^{\vee}) \times 0$ and $(\eta^{\vee}/H^{\vee}) \times 1$. Since f_t must preserve the intrinsic filtration of Y, it follows that f_t restricts to an isotopy of (Q, N). Thus $f_t((\xi^{\vee} \restriction K^{\vee}\langle K\rangle) \times I)$ has the properties ascribed to μ/P; this is a contradiction, which establishes our example. (A similar example, with more detailed proof, is given in Stone [24].)

<u>Theorem 3.1(n)</u>. Given polyhedra $X \subseteq Y$, a variety $\{\mathfrak{X}_n, \ldots, \mathfrak{X}_0\}$ of

X in Y, and $(Y^*, X^*) < \mathrm{bdy}(Y, X)$. Let $\underline{\xi}/\underline{K}$ and $\underline{\eta}/\underline{L}$ be $(n+1, n)$-stratifications of $\{\chi\}$ in Y with respect to (Y^*, X^*), such that $\underline{\xi}^* = \underline{\eta}^*$. (As usual, $\underline{\xi}^* = \{\beta \in \underline{\xi} : |\beta| \subset Y^*.\}$) Then there exist subdivisions $\underline{\xi}''/\underline{K}''$ of $\underline{\xi}$, $\underline{\eta}''/\underline{L}''$ of $\underline{\eta}$, and an isotopy h_t of (Y, X) such that $h_1 : \underline{\xi}'' \approx \underline{\eta}''$ is an isomorphism of stratifications, and h_t is the identity on Y^* for all t. (Of course, in the filtration (Y, X) it is understood that X has the filtration associated to $\{\chi\}$.)

The proof of Theorem 3.1 is by induction on n. Theorem 3.1(0) is a special case of the Appendix, Theorem I, 4.4. Now assume Theorem 3.1(n-1).

Lemma 3.2(n). Assume Theorem 3.1(n-1). Given an (n+1)-flag $(\xi_{n+1}, \ldots, \xi_1)/K$ such that for each $\sigma \in K$, $(\dot{\xi}_n(\sigma), \ldots, \dot{\xi}_1(\sigma))$ is a variety filtration of $\dot{\xi}_n(\sigma)$ in $\dot{\xi}_{n+1}(\sigma)$. Let $\underline{\eta}$ and $\underline{\zeta}$ be (n+1, n)-bs completions of (ξ) (as defined in Lemma 2.2). Then there are subdivisions $\underline{\eta}'$ of $\underline{\eta}$, $\underline{\zeta}'$ of $\underline{\zeta}$ and an isotopy f_t of $(|\xi|)$ such that:

$f_1 : \underline{\eta}' \approx \underline{\zeta}'$;

$\underline{\eta}'$ and $\underline{\zeta}'$ are (n+1, n)-bs completions of (ξ);

f_t is the identity on $|K|$ and a flag isomorphism of (ξ) for all t.

We use induction on a standard ordering of K. Assume that for every ρ preceding σ, $\Lambda^*(\rho, \underline{\eta}) = \Lambda^*(\rho, \underline{\zeta})$. Then $\Lambda^*(\sigma, \underline{\eta})$ and $\Lambda^*(\sigma, \underline{\zeta})$ are both (n, n-1)-stratifications of $(\dot{\xi}_n(\sigma))$ in $\dot{\xi}_{n+1}(\sigma)$, and they agree on $(\dot{\xi}_n(\partial\sigma))$ in $\dot{\xi}_{n+1}(\partial\sigma)$. By Theorem 3.1(n-1), there are subdivisions $\sigma\underline{\eta}'$ of $\Lambda^*(\sigma, \underline{\eta})$, $\sigma\underline{\zeta}'$ of $\Lambda^*(\sigma, \underline{\zeta})$ and an isotopy g_t of

$(\xi^{\bullet}(\sigma))$ rel $(\xi^{\bullet}(\partial\sigma))$ such that $g_1 : \sigma\underline{\eta}' \approx \sigma\underline{\xi}'$. Take $\underline{\eta}' = \underline{\eta} + \sigma\underline{\eta}'$, $\underline{\xi}' = \underline{\xi} + \sigma\underline{\xi}'$. It remains to extend g_t to an isotopy f_t of $(|\xi|)$. This is done using our standard ordering of K again. If ρ precedes σ, let f_t be the identity on $(|\xi(\sigma)|)$. Then f_t is defined on $(|\xi(\partial\sigma)|) \cup (\xi^{\bullet}(\sigma))$, so we define f_t on $(|\xi(\sigma)|)$ by conical extension to an interior point $\hat{\sigma}$ of σ; this leaves σ fixed. If τ suceeds σ, then f_t is inductively defined on $(|\xi(\partial\tau)|)$. Using a collar neighbourhood of $((|\xi(\partial\tau)|), (\xi^{\bullet}(\partial\tau)))$ in $((|\xi(\tau)|), (\xi^{\bullet}(\tau)))$ we extend f_t over $(|\xi(\tau)|)$; then f_t is the identity on τ and on the complement of the collar. Note that we may choose f_t to be the identity on $(|\xi(\tau)|)$ unless $\sigma < \tau$. Then f_t is a composition of finitely many local isotopies, so we have constructed a genuine p.ℓ. isotopy. The construction of f_t from g_t, and similarly inductively defined constructions, we shall refer to as the <u>standard procedure</u> for extending g_t.

Our inductive step is now complete. Since it uses a local construction, and since our block systems are locally finite, it follows that the (perhaps infinitely many) inductive steps do converge, and so we have proved Lemma 3.2(n), assuming Theorem 3.1(n-1).

<u>Corollary.</u> Assume Theorem 3.1(n-1). Given an (n+1, n)-stratification $\underline{\xi}/\underline{K}$ of X in Y. Let J be a cell-complex subdivision of $K_{o,o}$. Then there is a subdivision $\underline{\xi}'/\underline{K}'$ such that $K'_{o,o} = J$.

Let $(\xi'', K''_{,o})/J$ be any flag subdivision of $(\xi, K_{,o})/K_{o,o}$, and let $\underline{\eta}''/\underline{L}''$ be an (n+1, n)-bs completion of $(\xi'', K''_{,o})$, as constructed in

Lemma 2.2. Then for each $\sigma \in K_{o,o}$, $\Lambda^*(\sigma, \underline{\eta}'')$ is an $(n, n-1)$-stratification of $(K^{\bullet}_{,o}(\sigma))$ in $\xi^{\bullet}(\sigma)$: For the proof of Lemma 2.2, using a second-derived subdivision of a simplicial triangulation, shows that $|\Lambda^*(\sigma, \underline{\eta}'')|$ is a normal $(n, n-1)$-ns; and the axioms for a blocking are easily checked. (We could have interpolated Proposition 3.6 here.) Hence we have an $(n+1, n)$-bs completion of $(\xi, K_{,o})/K_{o,o}$ defined by: $\underline{\eta}^{\wedge}/\underline{L}^{\wedge} = (\xi, K_{,o}) \cup \Lambda^*_o(\underline{\eta}'')$. By the Lemma we can subdivide $\underline{\eta}^{\wedge}$ to another $(n+1, n)$-bs completion $\underline{\eta}^{\vee}/\underline{L}^{\vee}$ of $(\xi, K_{,o})$ and perform an isotopy f_t of $\underline{\eta}^{\vee}$ through completions of $(\xi, K_{,o})$ so that $f_1\underline{\eta}^{\vee}$ is a subdivision of $\Lambda_o(\underline{\xi})$. Now $\underline{\eta}'' + \Lambda^*_o(\underline{\eta}^{\vee})$ is a completion of $(\xi'', K''_{,o})/J$. Set $\Lambda_o(\underline{\xi}') = f_1(\underline{\eta}'' + \Lambda^*_o(\underline{\eta}^{\vee}))$; then $\Lambda_o(\underline{\xi}')$ is a subdivision of $\Lambda_o(\underline{\xi})$ with $K'_{o,o} = J$. So $\Lambda_o(\underline{\xi}') \cup (\mathcal{C}_o(\underline{\xi}) + \Lambda^*_o(\underline{\xi}'))$ is the required subdivision of $\underline{\xi}$. This proves the Corollary, still assuming Theorem 3.1(n-1).

Proof of Theorem 3.1(n): We are assuming Theorem 3.1(n-1) and hence Lemma 3.2(n). For the purposes of this proof we shall say an isotopy a_t of (Y, X) is "allowable" if a_t is the identity on Y^*. By means of an allowable isotopy, we may assume that the regular neighbourhoods $(|\xi_{n+1,o}|, \ldots, |\xi_{o,o}|)$ and $(|\eta_{n+1,o}|, \ldots, |\eta_{o,o}|)$ of X_o in (Y, X) are equal. The cell complexes $K_{o,o}$ and $L_{o,o}$ have a common subdivision rel $K^*_{o,o}$; call it H. By the Corollary to Lemma 3.2(n), $\underline{\xi}$ has a subdivision $\underline{\xi}^{\sim}/\underline{K}^{\sim}$ with $K^{\sim}_{o,o} = H$; similarly let $\underline{\eta}^{\sim}/\underline{L}^{\sim}$ subdivide $\underline{\eta}$ with $L^{\sim}_{o,o} = H$. It follows from the proofs of Lemmas 2.2 and 3.2(n) that we may assume that $\underline{\xi}^{\sim *}$ (defined as $\underline{\xi}^{\sim} < \underline{\xi}^{*} >$) and $\underline{\eta}^{\sim *}$

are equal. The flags $(\xi^{\sim}, K^{\sim}_{,o})/K^{\sim}_{o,o}$ and $(\eta^{\sim}, L^{\sim}_{,o})/L^{\sim}_{o,o}$ have the same sequence of underlying polyhedra, the same rim, the same cell complex H for base, and they are equal over $K^*_{o,o} < \text{bdy } H$. It follows from the Appendix, Theorem I, 4.4 that there is an allowable isotopy b_t such that:

$b_1 : (\xi^{\sim}, K^{\sim}_{,o}) \approx (\eta^{\sim}, L^{\sim}_{,o})$ is an isomorphism of flags;

b_t restricts to isotopies of $(|\xi|, |K_{,o}|)$ and $(\xi^{\bullet}, K^{\bullet}_{,o})$.

Henceforth I shall suppress b_t and assume the flags $(\xi^{\sim}, K^{\sim}_{,o})$ and $(\eta^{\sim}, L^{\sim}_{,o})$ are equal.

By Lemma 3.2(n), after subdividing $\Lambda^*_o(\underline{\xi}^{\sim})$ and $\Lambda^*_o(\underline{\eta}^{\sim})$ if necessary (thus inducing subdivisions of $\underline{\xi}^{\sim}$ and $\underline{\eta}^{\sim}$), there is an isotopy c_t of $(|\Lambda_o(\underline{\xi})|)$ such that $c_1 : \Lambda^*_o(\underline{\xi}^{\sim}) \approx \Lambda^*_o(\underline{\eta}^{\sim})$ and c_t is a flag isomorphism of $(\xi^{\sim}, K^{\sim}_{,o}) = (\eta^{\sim}, L^{\sim}_{,o})$ for all t. The proof of the Lemma shows we may choose c_t to be the identity on $(|\Lambda_o(\underline{\xi}^*)|)$, since $\Lambda^*_o(\underline{\xi}^{\sim})$ and $\Lambda^*_o(\underline{\eta}^{\sim})$ are already equal there. Since $(|\Lambda^*_o(\underline{\xi}^{\sim})| \cup |\mathcal{C}_o(\underline{\xi}^*)|)$ $< \text{bdy } (|\mathcal{C}_o(\underline{\xi})|)$, c_t extends to an allowable isotopy f_t of (Y, X).

Now set $Y' = \text{cl}[Y - |\eta_{q,o}|]$, and let (Y', X') be the restriction of (Y, X). Take $Y'^* = (Y^* \cap Y') \cup \eta^{\bullet}_{q,o}$; then $(Y'^*, X'^*) < \text{bdy}(Y', X')$. Both $f_1(\mathcal{C}_o\underline{\xi}^{\sim})$ and $\mathcal{C}_o\underline{\eta}^{\sim}$ are $(n, n-1)$-stratifications of (X') in Y' with respect to Y'^*; and their restrictions to Y'^* are respectively $f_1(\mathcal{C}_o\underline{\xi}^{\sim *} \cup \Lambda^*_o\underline{\xi}^{\sim}) = \mathcal{C}_o\underline{\xi}^* \cup f_1(\Lambda^*_o\underline{\xi}^{\sim})$ and $\mathcal{C}_o\underline{\eta}^{\sim} \cup \Lambda^*_o\underline{\eta}^{\sim} = \mathcal{C}_o\underline{\eta}^* \cup \Lambda^*_o\underline{\eta}^{\sim}$, which are equal. So by inductive hypothesis, Theorem 3.1(n-1), there are subdivisions $\underline{\xi}^{\wedge}/\underline{K}^{\wedge}$ of $f_1(\mathcal{C}_o\underline{\xi}^{\sim})$, $\underline{\eta}^{\wedge}/\underline{L}^{\wedge}$ of $\mathcal{C}_o\underline{\eta}^{\sim}$, and an isotopy g_t of (Y', X') such that $g_1 : \underline{\xi}^{\wedge} \approx \underline{\eta}^{\wedge}$ and g_t is the identity on Y'^*. g_t extends to an allowable isotopy k_t of Y by the identity on $|\eta_{q,o}|$. Then the subdivision $\underline{\xi}''/\underline{K}'' = \underline{\xi}^{\sim} + \underline{\eta}^{\wedge}$ of $\underline{\eta}$, and the isotopy $h_t = k_t \circ f_t$ of Y satisfy the requirements of Theorem 3.1(n).

Thus Theorem 3.1(n) and with it, Lemma 3.2(n), hold for all n.

Corollary 1. If, in addition to the hypotheses of Theorem 3.1(n), we are given that the (n+1, n)-ns's $|\underline{\xi}|$ and $|\underline{\eta}|$ are equal, then we may choose h_t to respect this ns; that is, so that $h_t |\xi_{j,i}| = |\xi_{j,i}|$ for all $0 \leq i \leq n$, $i \leq j \leq n+1$.

This is proved by induction on n together with Theorem 3.1(n). This leaves our isotopy b_t that need not respect the ns $|\underline{\xi}|$. One first chooses b_t to respect the (n, n-1)-ns $|\Lambda_o^*(\underline{\xi})|$ by an inductively applied argument using the uniqueness of normal ns's and the standard procedure. Now a collaring of $|\Lambda_o^*(\underline{\xi})|$ in $|\mathcal{C}_o(\underline{\xi})|$ shows that one can choose b_t to respect the (n, n-1)-ns $|\mathcal{C}_o(\underline{\xi})|$. Details are left to the reader.

Corollary 2. Let $\underline{N}$ be an (n+1, n)-ns, $\underline{N}^* < \text{bdy } \underline{N}$. Assume of the associated filtered polyhedron $(N_{n+1,}, \ldots, N_{o,})$ that $(N_{n,}, \ldots, N_{o,})$ is a variety filtration of $N_{n,}$ in $N_{n+1,}$. Given blockings $\underline{\xi}/\underline{K}$ and $\underline{\eta}/\underline{L}$ of $\underline{N}$ such that $\underline{\xi}^* = \underline{\eta}^*$. Then there exist subdivisions $\underline{\xi}''/\underline{K}''$ of $\underline{\xi}$, $\underline{\eta}''/\underline{L}''$ of $\underline{\eta}$ and an isotopy f_t of the ns $\underline{N}$ which is fixed on $\underline{N}^*$, such that $f_1 : \underline{\xi}'' \approx \underline{\eta}''$.

Define $Y = N_{n+1,} \cup \text{fr } N_{n+1,} \times I$ to be the disjoint union with $\text{fr } N_{n+1,}$ identified to $\text{fr } N_{n+1,} \times 0$ in the obvious way. Then $\underline{\xi}$ and $\underline{\eta}$

are $(n+1, n)$-stratifications of $(N_{n,}\ )$ in Y with respect to $Y^* = N^*_{n+1,} \cup \mathrm{fr}\, N^*_{n+1,} \times I \subseteq Y$. So Corollary 2 follows from Corollary 1.

Corollary 3. If, in addition to the hypotheses of Theorem 3.1(n), we are given that the stratifications $\underline{K}$ and $\underline{L}$ of X are equal, then we may choose $\underline{\xi}'' = \underline{\xi}$, $\underline{\eta}'' = \underline{\eta}$ and h_t to be the identity on X.

This is proved by induction on n together with Theorem 3.1(n). We no longer need Lemma 3.2(n); however we do need the full force of the Appendix, Theorem I, 4.4 to ensure that b_t is the identity on X. Other details are left to the reader.

Corollary 4. If, in addition to the hypotheses of Theorem 3.1(n), we are given that the blocks of $\underline{\xi}$ and $\underline{\eta}$ are δ-small, then we may require h_t to be an ϵ-isotopy. Here δ depends on ϵ, n, and the dimension of Y (but not on the stratifications $\underline{\xi}$ or $\underline{\eta}$).

Again we follow the proof of Theorem 3.1(n), and show that all isotopies may be made small. Note that the inductively constructed adjustment of b_t to f_t preserves the blocks of $(\eta, L_{,o})/L_{o,o}$. Hence this adjustment must be small on $|(\eta, L_{,o})|$, and so may be extended to a δ- small allowable isotopy of Y. Thus f_t is a δ-isotopy. The rest of the proof is left to the reader.

Corollary 5. One can replace "Y" in hypotheses and conclusion of Theorem 3.1 by "a filtered polyhedron $(Y_q, \ldots, Y_o)$", and "$X \subseteq Y$" by "$X \subseteq Y_o$"; then the modified statement is true.

The proof is left to the reader.

As an application of the techniques used so far, we shall now prove a product theorem for stratifications: given a stratification ξ whose base space is of the form $\underline{K} \times I$, then $\underline{\xi}$ itself is of the form $\underline{\eta} \times I$, where $\underline{\eta}$ is a stratification over $\underline{K}$. We have first to define the "product" of a stratification with the interval I. Let me point out that this theorem accords with our notion of $(n+1, n)$-stratification as generalized bundles over n-stratifications, since the product theorem (has meaning and) is true in all regular bundle theories (so far as I know).

After the product theorem we shall return to our program of proving uniqueness theorems. Having hitherto considered separately each pair $X \subset Y$ with a variety of X in Y, we shall then consider simultaneously all pairs $X \subseteq Y$ for fixed X and variety of X.

Let β and γ be blocks. Then their cartesian <u>product</u> is the block $\beta \times \gamma$ defined by: $(|\beta| \times |\gamma|, \{\ell\beta \times \ell\gamma, \partial\beta \times |\gamma| \cup |\beta| \times \partial\gamma, \beta^{\bullet} \times |\gamma| \cup |\beta| \times \gamma^{\bullet})$
If β has fibre F and γ has fibre G, then $\beta \times \gamma$ has for fibre the join $F*G$. In particular if β and γ are cells, that is, disks with the canonical block structure, then $\beta \times \gamma$ is a cell.

Given $(X) = (X_n, \ldots, X_o)$, $(Y) = (Y_m, \ldots, Y_o)$; then $X \times Y$ has the $(m+n)$-filtration defined by:

$(X \times Y)_k = \cup\{X_i \times Y_j : i + j = k\}$, for $k = 0, \ldots, m+n$.

Let $\underline{N}, \underline{M}$ be normal (n, n)- and (m, m)-rns's for (X) and (Y). Set

$(N \times M)_{\ell, k} = \cup\{N_{j,i} \times M_{h,g} : i + g = k, \; j + h = \ell\}$,

for $0 \leq k \leq \ell \leq m+n$. This defines a family of polyhedra $\underline{N \times M}$. In general $\underline{N \times M}$ is not any kind of ns. For though $((N \times M)_{,o})$ is a regular neighbourhood of $X_o \times Y_o$ in $(X \times Y)$, $\{(N \times M)_{\ell,k} \cap ((N \times M)^{\bullet}_{,o})\}$ is not a normal rns in general.

Lemma 3.3. 1. Let N be a (q, n)-ns, $\underline{N}^* < \text{bdy } \underline{N}$. Given a polyhedron Z, and $Z^* < \text{bdy } Z$. Then

$\underline{N \times Z} = \{N_{j,i} \times Z\}$ is a (q, n)-ns, and

$\underline{N \times Z^*} = \{N_{j,i} \times Z^* \cup N^*_{j,i} \times Z\}$ is $< \text{bdy } \underline{N \times Z}$.

2. If $\underline{\xi}$ is a blocking of $\underline{N}$ with respect to $\underline{N}^*$, and K is a cell complex triangulation of Z in which Z^* is covered by a subcomplex $K^* < K$, then $\underline{\xi} \times B = \{\beta \times \tau : \beta \in \underline{\xi},\ \tau \in K\}$ is a blocking of $\underline{N \times Z}$ with respect to $\underline{N \times Z^*}$.

Corollary 1. For any n-sheaf $\underline{\xi}/\underline{K}$, there is defined the n-sheaf $\underline{\xi} \times I/\underline{K} \times I$, also denoted $(\underline{\xi}/\underline{K}) \times I$. Here $I = [0, 1]$ is regarded as the standard 1-simplex.

Corollary 2. Let K be a cell complex, $(F_n, \ldots, F_1)$ a compact filtered polyhedron. Then we have the trivial $c(F_n, \ldots, F_1)$-block bundle $(\xi)/K$ defined by: $(|\xi_n|, \ldots, |\xi_1|) = |K| \times c(F_n, \ldots, F_1)$, and $\xi_i = \{\sigma \times cF_i : \sigma \in K\}$, for $i = 1, \ldots, n$. (Here cF_i is regarded as having the block structure $(cF_i, \{c, \emptyset, F_i\})$.) We shall usually denote $(\xi)/K$ by: $K \times c(F)$. A trivialization of a c(F)-bb $(\eta)/K$ is a flag isomorphism $f : (\eta) \approx K \times c(F)$ modulo K. Since any n-flag is a disjoint union of $c(F^i)$-block bundles over the components of K, we can define a trivialization of any n-flag.

Proposition 3.4. Let $\underline{K}$ be an n-bs, $\underline{\xi}$, $\underline{\eta}$ (n+1, n)-bs's over $\underline{K}$. Then any bs isomorphism modulo $\underline{K} \times 0$ $p : \underline{\xi} \restriction \underline{K} \times 0 \approx \underline{\eta} \restriction \underline{K} \times 0$ extends to an isomorphism modulo $\underline{K} \times I$ $p'' : \underline{\xi} \approx \underline{\eta}$. Further if $\underline{K}' < \underline{K}$ and we are given an isomorphism modulo $\underline{K}' \times I$ $p' : \underline{\xi} \restriction \underline{K}' \times I \approx \underline{\eta} \restriction \underline{K}' \times I$ which extends p, then we may take p'' to extend p'.

Corollary. Any (n+1, n)-bs $\underline{\xi}/\underline{K} \times I$ is isomorphic modulo $\underline{K} \times I$ to $(\underline{\xi} \restriction \underline{K} \times 0) \times I$.

The proof of the proposition is an inductive application of this lemma:

Lemma 3.5. Let $\underline{\xi}$ and $\underline{\eta}$ be weak (n+1, n)-stratifications over a weak n-stratification $\underline{K}$ such that $K_{o,o}$ is a single q-cell $<\sigma^q>$. Let ρ^{q-1} be a (q-1)-face of σ, and set $H = K_{o,o} - \{\rho, \sigma\}$ (thus $H < K_{o,o}$ is a sub-complex), $\underline{K}' = \Lambda(H, \underline{K}) \cup \Lambda^*(\sigma, \underline{K})$. Given an isomorphism modulo $\underline{K}'$ $p : \underline{\xi} \restriction \underline{K}' \approx \underline{\tau} \restriction \underline{K}'$. Then p extends to an isomorphism $p'' : \underline{\xi} \approx \underline{\tau}$ modulo $\underline{K}$.

By applying Lemma 1.7 to a vertex of H we see that the flags $(\xi, K_{,o})/<\sigma>$ and $(\eta, K_{,o})/<\sigma>$ have the same fibre, say $(G, F_n, \ldots, F_1)$. Since σ is contractible, these flags are trivial (see the Appendix, Theorem I, 1.1). Let $h : (\xi, K_{,o})/<\sigma> \to <\sigma> \times c(G, F)$ be a trivialization. Since $(F_n, \ldots, F_1)$ is a variety filtration of $F = F_n$ in G, by bs 3, we may take an (n, n-1)-stratification $\mathfrak{h}^\wedge/\underline{L}^\wedge$ of (F) in G, by Theorem 2.1. Then there is a weak (n+1, n)-stratification $\mathfrak{h}/\underline{L}$ on $c(F)$ in cG defined by $\mathfrak{h} = \mathfrak{h}^\wedge \cup \{cG, cF_n, \ldots, cF_1, c\}$. (Here cG is given the block structure $(cG, \{c, \emptyset, G\})$. Hence we have a weak (n+1, n)-stratification

$\not{\xi} \times \langle\sigma\rangle / \underline{L} \times \langle\sigma\rangle$ on $\sigma \times c(G, F)$; and since the p.ℓ. isomorphism $h : |\underline{\xi}| \to |\not{\xi} \times \langle\sigma\rangle|$ is an isomorphism of flags $h : (\xi_{,o}) \approx ((\not{\xi}\times\langle\sigma\rangle)_{,o})$, the proof of Lemma 3.2 shows that we may find subdivisions $\underline{\xi}'$ of $\underline{\xi}$, $(\not{\xi}\times\langle\sigma\rangle)'$ of $\not{\xi} \times \langle\sigma\rangle$, and an isotopy h_t of h such that $h_1 : \underline{\xi}' \approx (\not{\xi}\times\langle\sigma\rangle)'$ is an isomorphism of stratifications. Now the p.ℓ. isomorphism $h_1 : |\underline{\xi}| \to \sigma \times c(G, F)$ corresponds

$|\xi_{n,o} \restriction H| \quad \longleftrightarrow |H| \times cG;$

$|\xi_{n,o}(\rho)| \quad \longleftrightarrow \rho \times cG;$

$|\underline{\xi} \restriction \Lambda^*(\sigma, \underline{K})| \longleftrightarrow \sigma \times V$, where $V = |\not{\xi}^{\wedge}|$ is a regular neighbourhood of F in G;

$\mathrm{fr}\, \underline{\xi} \restriction \Lambda^*(\sigma, \underline{K}) \longleftrightarrow \sigma \times \mathrm{fr}\, V.$

Let $k_1 : |\underline{\eta}| \to \sigma \times c(G, F)$ be an analogous p.ℓ. isomorphism. Then $f = k_1 \circ p \circ h_1^{-1}$ is a p.ℓ. automorphism of $|H| \times cG \cup \sigma \times V$, which restricts to automorphisms of $\sigma \times \mathrm{fr}\, V$, $\rho \times V$, $\partial\rho \times cG$ and $|H| \times \mathrm{cl}[G-V]$. Now

$$\begin{aligned}\partial\sigma \times cG \cup \sigma \times G = &(H \times cG \cup \sigma \times V)\\ &\cup (\rho \times cG \cup \sigma \times \mathrm{cl}[G-V]).\end{aligned}$$

Let $\hat{\rho}$ be an interior point of ρ; then $\rho \times cG$ is a cone from $\hat{\rho}$ to $\partial\rho \times cG \cup \rho \times G$. It follows that $\rho \times cG \cup \sigma \times \mathrm{cl}[G-V]$ is a cone from $\hat{\rho}$ to $\sigma \times \mathrm{fr}\, V \cup \rho \times V \cup \partial\rho \times cG \cup |H| \times \mathrm{cl}[G-V]$ (see diagram 7, p. 3.14). So f extends to a p.ℓ. automorphism f' of $\partial\sigma \times cG \cup \sigma \times G$, by extending conically to $\hat{\rho}$. Moreover, both $\rho \times cG$ and $f'(\rho \times cG)$ are regular neighbourhoods of $\rho \times (cF \cup V) \cup \partial\rho \times cG$ rel $\rho \times \mathrm{fr}\, V \cup \cup\, \partial\rho \times G$ in $\rho \times cG \cup \sigma \times \mathrm{cl}[G-V]$. Hence we may choose f' so that

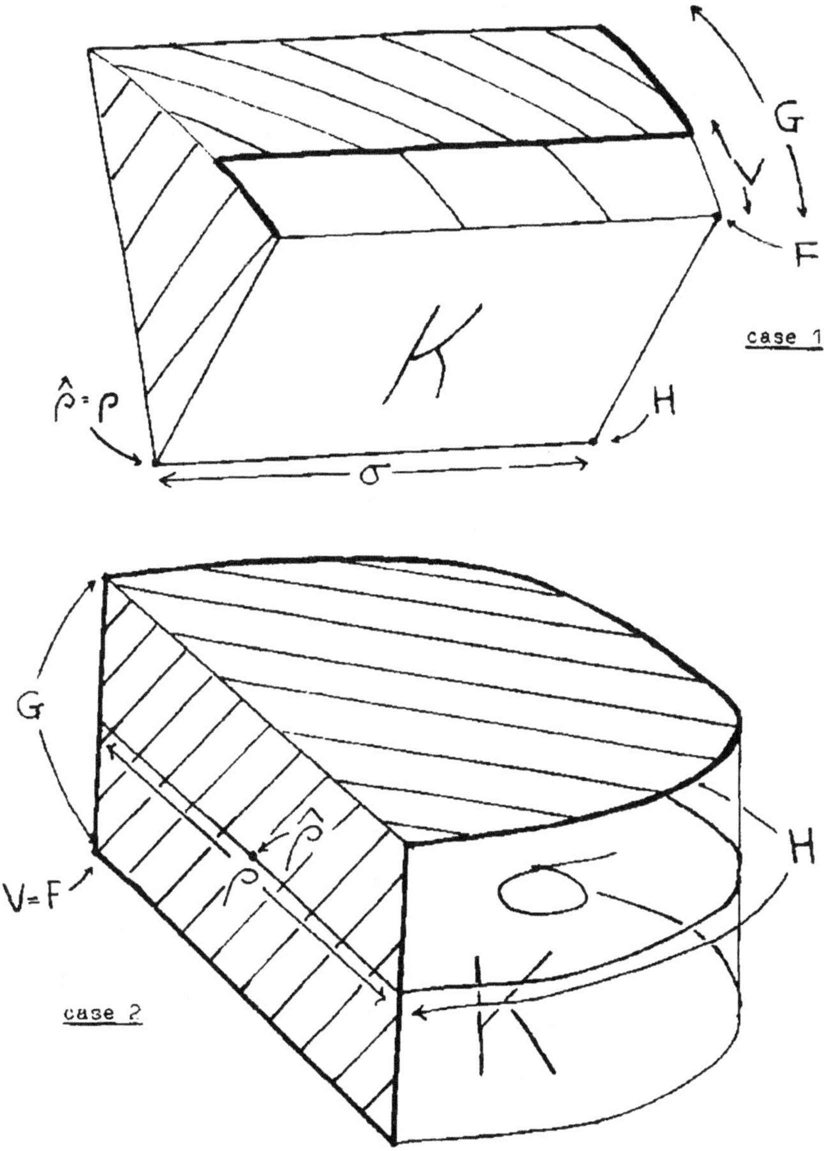

diagram 7

$f''(\rho \times cG) = \rho \times cG$. Take a point $\hat{\sigma}$ in the interior of σ; then $\sigma \times cG$ is a cone from $\hat{\sigma}$ to $\partial\sigma \times cG \cup \sigma \times G$. So f'' extends to a p.ℓ. automorphism f' of $\sigma \times cG$ by extending conically to $\hat{\sigma}$. Observe that f' respects the variety filtration $\sigma \times c(F_n, \ldots, F_1, \emptyset)$ of $\sigma \times cF$ in $\sigma \times cG$; and that f'' extends over $\sigma \times cF$ by the automorphism $k_1 \circ h_1^{-1}$, which also respects this variety filtration. By Akin's Cone Unknotting Theorem [1, Corollary IV. 8] (see also our quoted Theorem 1.6, 5), we may choose f' to extend $f'' \cup k_1 \circ h_1^{-1}$. Then $p'' = k_1^{-1} \circ f' \circ h_1$ is an isomorphism of weak stratifications, $p'' : \underline{\xi} \approx \underline{\eta}$ modulo $\underline{K}$.

Proof of Proposition 3.4: Take a standard ordering of the cells of $\underline{K}$. We use induction on this ordering to construct $p'' : \Lambda(\sigma \times I, \underline{\xi}) \approx \Lambda(\sigma \times I, \underline{\eta})$ so as to extend p, p' and the previously defined p''; for p'' is already defined on $\Lambda(\partial\sigma \times I \cup \sigma \times 0, \underline{\xi}) \cup \Lambda^*(\sigma \times I, \underline{\xi})$. If $\sigma \in \underline{K}'$, we extend p'' by p'; otherwise we apply Lemma 3.5 to the weak stratifications $\Lambda(\sigma \times I, \underline{\xi})$ and $\Lambda(\sigma \times I, \underline{\eta})$, and the face $\sigma \times 1$ of $\sigma \times I$.

We now return to our program of proving uniqueness theorems. Consider our stock question: given X and a variety $\{\mathfrak{X}\}$ of X, how many abstract regular neighbourhoods Y of X are there such that $\{\mathfrak{X}\}$ is a variety of X in Y? It will be convenient, at least conceptually, to be able to fix a stratification $\underline{K}$ of $\{\mathfrak{X}\}$ and describe each Y by a stratification $\underline{\xi}$ over this fixed $\underline{K}$. This also has the advantage that, for each Y, $\underline{\xi}$ is unique up to isotopy of Y rel X - a subdivision and isotopy of (Y, X) are unnecessary. Now let $\underline{\xi}'/\underline{K}'$ be some stratification of $\{\mathfrak{X}\}$ in Y.

We know there is a subdivision $\underline{K}''$ of $\underline{K}'$ and an isotopy f_t of (X) such that $f_1\underline{K}''$ is a subdivision of $\underline{K}$. Thus we have first to subdivide $\underline{\xi}'$ to some $\underline{\xi}''/\underline{K}''$. The isotopy f_t of X extends to an isotopy g_t of Y; but then we have to form from the stratification $g_1\underline{\xi}''/f_1\underline{K}''$ of $\{X\}$ in Y a stratification $\underline{\xi}$ over $\underline{K}$. This process of "amalgamation" is the new step. It is almost trivial, once we know the non-trivial fact that disk block bundles over cell complexes can be amalgamated (see Rourke and Sanderson [19, I]).

Proposition 3.6. Let $\underline{K}$ be an n-bs, $\underline{L}$ a subdivision of $\underline{K}$. Given an (n+1, n)-bs $\underline{\eta}/\underline{L}$, then there is a unique (n+1, n)-bs $\underline{\xi}/\underline{K}$ of which $\underline{\eta}$ is a subdivision.

$\underline{\xi}$ is called the amalgamation of $\underline{\eta}$ over $\underline{K}$.

The proof also applies to a (q, n)-bs $\underline{\eta}$ whose n-bs is $\underline{L}$; one constructs an amalgamated (q, n)-bs $\underline{\xi}$ whose n-bs is $\underline{K}$.

For each $i = 0, \ldots, n$, let $(\xi, K_{,i})/K_{i,i}$ be the flag amalgamation of $(\eta, L_{,i})/L_{i,i}$; that is, the unique flag over $K_{i,i}$ of which $(\eta, L_{,i})$ is a subdivision (see the Appendix, section I, 1). Then $\underline{\xi}$ has to be the set of blocks $\cup\{(\xi, K_{,i}) : i = 0, \ldots, n\}$; we must show that $\underline{\xi}$ is an (n+1, n)-bs. $\{|\xi_{j,i}|\}$ forms an (n+1, n)-ns, since it is the same as $\{|\eta_{j,i}|\}$. The intersection condition bs 4 for $\underline{\xi}$ to be a blocking is straight-forward to check. We have to prove bs 3: that for every cell $\sigma \in K_{k,k'}$ $\{|\xi(\sigma)| \cap |\eta_{j,i}|\}$ is an (n+1-k, n-k)-ns for $(|K_{,k}(\sigma)|)$ in $|\xi(\sigma)|$, and that $\{\beta \in \underline{\xi} : |\beta| \subseteq |\xi(\sigma)|\}$ is a blocking of this ns. This reduces to showing that $\Lambda^*(\sigma, \underline{\xi})$ is an (n+1-k, n-k)-stratification for

$(K^{\bullet}_{,k})(\sigma)$ in $\underline{\xi}^{\bullet}(\sigma)$ (where the set of blocks $\Lambda^*(\sigma,\underline{\xi})$ is defined in the obvious way). By decreasing induction on k, we may assume that $\Lambda^*(\pi,\underline{\xi})$ is an $(n+1-\ell, n-\ell)$-stratification for $(K^{\bullet}_{,\ell})(\pi)$ in $\underline{\xi}^{\bullet}(\pi)$ whenever $\pi \in K_{\ell,\ell}$ with $\ell \underset{+}{>} k$. Take a standard ordering $\tau_1,\ldots,\tau_s,\ldots$ of $\{\tau \in L_{k,k} : \tau \subseteq \sigma\}$. By induction on r one shows that $\cup\{|\Lambda^*(\tau_s,\underline{\eta})| : s = 1,\ldots,r\}$ is a normal $(n+1-k, n-k)$-ns for $\cup\{(L^{\bullet}_{,k})(\tau_s)\}$ in $\cup\{\eta^{\bullet}(\tau_s)\}$; the inductive step follows from Lemma 1.12. At the end of the induction we have that $|\Lambda^*(\sigma,\underline{\xi})|$ is a normal $(n+1-k, n-k)$-ns for $(K^{\bullet}_{,k})(\sigma)$ in $\underline{\xi}^{\bullet}(\sigma)$. As before, $\Lambda^*(\sigma,\underline{\xi})$ satisfies <u>bs 1</u>, <u>2</u> and <u>4</u>; and <u>bs 3</u> holds by inductive hypothesis, since every cell $\pi \in \Lambda^*(\sigma,\underline{\xi})$ is in some $K_{\ell,\ell}$ with $\ell \underset{+}{>} k$. So $\Lambda^*(\sigma,\underline{\xi})$ is a stratification, as required. The general step in our induction on k is complete, and Proposition 3.6 follows.

<u>Proposition 3.7(n)</u>. Given an $(n+1,n)$-bs $\underline{\xi}/\underline{K}$ and a subdivision $\underline{L}$ of $\underline{K}$. Then there is a subdivision $\underline{\eta}/\underline{L}$ of $\underline{\xi}$.

The proof also applies to a (q,n)-bs $\underline{\xi}$ whose n-bs is $\underline{K}$; one constructs a (q,n)-bs $\underline{\eta}$ subdividing $\underline{\xi}$ whose n-bs is $\underline{L}$.

We use induction on n. Proposition 3.7(0) is a special case of the Appendix, Theorem I, 1.5. Now assume Proposition 3.7(n-1):

There is a flag subdivision $(\mu, L_{,o})/L_{o,o}$ of $(\xi, K_{,o})/K_{o,o}$, by the Appendix, Theorem I, 1.5. Let $\underline{\mu}''/P''$ be a weak $(n+1,n)$-bs completion of $(\mu, L_{,o})/L_{o,o}$.

I claim that $\underline{P}''$ can be chosen to subdivide $\Lambda_o(\underline{L})$. For by

Lemma 3.2, there are a subdivision $\underline{P}'$ of $\Lambda_o^*(\underline{P}'')$ and an isotopy a_t of $(|L_{,o}|)$ such that $a_1\underline{P}'$ subdivides $\Lambda_o^*(\underline{L})$ and a_t is a flag isomorphism of $(L_{,o})$ for all t. By inductive hypothesis, Proposition 3.7(n-1), there is a subdivision $\underline{\mu}'/\underline{P}'$ of $\Lambda_o^*(\underline{\mu}'')$. We cannot directly extend a_t to an isotopy of $(|\xi_{,o}|)$, since $(|K_{,o}|)$ need not be a variety filtration ($\underline{\xi}$ is any (n+1, n)-bs). However, for each cell $\sigma \in K_{o,o}$, we have the restriction $\underline{\mu}''<\Lambda^*(\tau, \underline{\xi})>$, which is a stratification. Since a_t respects each $\underline{P}''<\Lambda^*(\sigma, \underline{K})>$, a_t extends to an isotopy of $|\underline{\mu}''<\Lambda^*(\sigma, \underline{\xi})>|$. Using a standard procedure, we can thus extend a_t to an isotopy b_t of $(|\xi_{,o}|)$ such that $b_t(\mu'', P''_{,o})$ is a flag subdivision of $(\xi, K_{,o})$ for all t. Then $b_1(\underline{\mu}''+\underline{\mu}')$ is still a completion of $(\mu, L_{,o})$, for some choice of μ, and $b_1(\underline{P}''+\underline{P}')$ subdivides $\Lambda_o(\underline{L})$. This justifies our claim.

Now that $\underline{P}''$ subdivides $\underline{L}$, $\underline{\mu}''$ can be amalgamated to a weak (n+1, n)-bs $\underline{\eta}''/\Lambda_o(\underline{L})$, by Proposition 3.6. By inductive hypothesis, Proposition 3.7(n-1), $\mathcal{C}_o(\underline{\xi})$ has a subdivision $\underline{\eta}'/\mathcal{C}_o(\underline{L})$. However, $\Lambda_o^*(\underline{\eta}'')$ and $\underline{\eta}'<\Lambda_o^*(\underline{\xi})>$ need not be equal.

Take a standard ordering of the cells of $K_{o,o}$, and assume inductively that we have chosen $\underline{\eta}''$ so that $\Lambda^*(\pi, \underline{\eta}'') = \underline{\eta}'<\Lambda^*(\pi, \underline{\xi})>$ whenever π precedes τ. Then $\Lambda^*(\tau, \underline{\eta}'')$ and $\underline{\eta}'<\Lambda^*(\tau, \underline{\xi})>$ are both (n, n-1)-stratifications of $(K^\bullet_{,o})(\tau)$ in $\xi^\bullet(\tau)$ with respect to $\xi^\bullet(\partial\tau)$ which have the same (n-1)-bs $\Lambda^*(\tau, \underline{L})$ and which agree in $\xi^\bullet(\partial\tau)$. By Corollary 3 to Theorem 3.1 there is an isotopy d_t of $(\xi^\bullet, K^\bullet_{,o})(\tau)$ such that $d_1 : \Lambda^*(\tau, \underline{\eta}'') \approx \underline{\eta}'<\Lambda^*(\tau, \underline{\xi})>$ and d_t is the identity on $(K^\bullet_{,o})(\tau)$ and on $\xi^\bullet(\partial\tau)$. Extend d_t to an isotopy e_t of $|(\xi, K_{,o})|$ by the standard procedure; then $e_1(\underline{\eta}'')$ is still a weak $(n+1, n)$-bs over $\Lambda_o(\underline{L})$, still subdivides $\Lambda_o(\underline{\xi})$, and $\Lambda^*(\pi, e_1\underline{\eta}'') = \underline{\eta}'<\Lambda^*(\pi, \underline{\xi})>$ whenever π precedes or equals τ.

At the end of the induction we have adjusted $\underline{\eta}''$ till $\Lambda_o^*(\underline{\eta}'') = \underline{\eta}'<\Lambda_o^*(\underline{\xi})>$. Since this latter is $<$ bdy $\underline{\eta}'$, we may form the $(n+1, n)$-bs $\underline{\eta}/\underline{L} = \underline{\eta}'' \cup \underline{\eta}'$, by Lemma 1.13; then $\underline{\eta}$ is the required subdivision of $\underline{\xi}$.

Remark. One can show that given an $(n+1, n)$-bs $\underline{\xi}/\underline{K}$ and a subdivision $\underline{L}$ of $\underline{K}$, if $\underline{\eta}/\underline{L}$ and $\underline{\zeta}/\underline{L}$ are subdivisions of $\underline{\xi}$, then there is an isotopy f_t of $|\underline{\xi}|$ such that $f_1 : \underline{\eta} \approx \underline{\zeta}$, f_t is the identity on $|\underline{K}|$, and f_t respects the blocks of $\underline{\xi}$ (that is, $f_t : \underline{\xi} \approx \underline{\xi}$ is a bs isomorphism) for all t. Thus in this context, subdivision is unique up to isotopy. One might conjecture a more general result: given $\underline{\xi}/\underline{K}$ and subdivisions $\underline{\eta}/\underline{L}$, $\underline{\zeta}/\underline{H}$ of $\underline{\xi}$; then there should exist subdivisions $\underline{\eta}'/\underline{L}'$ of $\underline{\eta}$, $\underline{\zeta}'/\underline{H}'$ of $\underline{\zeta}$, and an isotopy f_t of $|\underline{\xi}|$ such that $f_1 : \underline{\eta}' \approx \underline{\zeta}'$, and f_t respects the blocks of $\underline{\xi}$ for all t. This conjecture is false in general even for stratifications. (See diagram 8, p. 3.20). The best that can be done in this context is that subdivision is unique, not up to isotopy, but up to "concordance", a notion which we define now:

Lemma 3.8. Let $\underline{\xi}/\underline{K}$ be an $(n+1, n)$-stratification. Let $\underline{\eta}/\underline{L}$ and $\underline{\zeta}/\underline{H}$ be subdivisions of $\underline{\xi}$. Then there exist a subdivision $(\underline{\xi} \times I)''/(\underline{K} \times I)''$ of $\underline{\xi} \times I/\underline{K} \times I$, and isotopies f_t, g_t of the $(n+1, n)$-ns $|\underline{\xi} \times I|$ such that:

$f_1 : (\underline{\xi} \times I)'' \to \underline{\eta} \times I$ and $g_1 : (\underline{\xi} \times I)'' \to \underline{\zeta} \times I$ are subdivisions;

f_t is the identity on $|\underline{\xi} \times 0|$ for all t, and g_t is the identity on $|\underline{\xi} \times 1|$ for

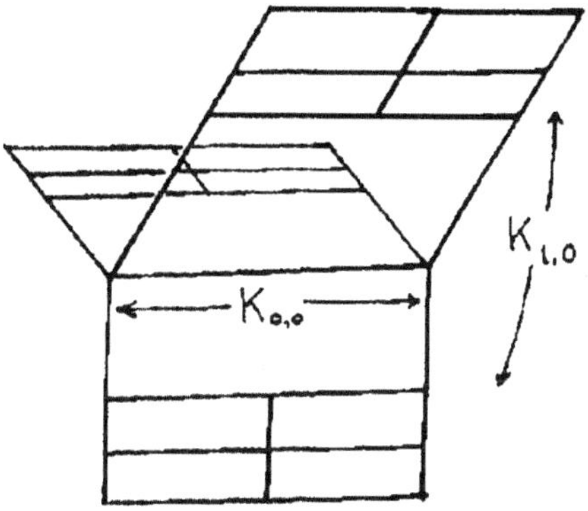

A 1-bs structure $\underline{K}$ on the space "Y" x I

Subdivisions $\underline{L}$ and $\underline{H}$ of $\underline{K}$

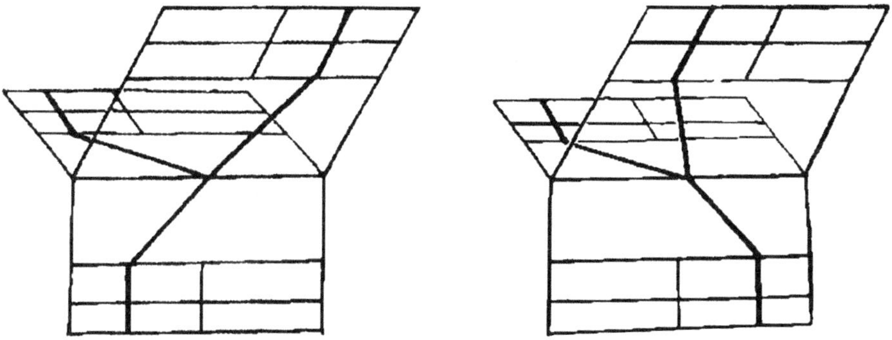

$\underline{L}$ $\underline{H}$

No subdivisions of $\underline{L}$ and $\underline{H}$ are isotopic through subdivisions of $\underline{K}$.

diagram 8

all t.

The structure $((\underline{\xi}\times I)'', f_t, g_t)$ will be called a <u>concordance</u> between $\underline{\eta}$ and $\underline{\zeta}$.

Set $\underline{\mu}/\underline{M} = (\underline{\xi}\times I+\underline{\eta}\times 0+\underline{\zeta}\times 1)/(\underline{K}\times I+\underline{L}\times 0+\underline{H}\times 1)$. Then $\underline{\mu}$ and $\underline{\eta}\times I$ are both blockings of the ns $|\underline{\xi}\times I|$ which agree in $|\underline{\xi}\times 0| < \mathrm{bdy}|\underline{\xi}\times I|$. By Corollary 2 to Theorem 3.1, there are a subdivision $\underline{\mu}'/\underline{M}'$ of $\underline{\mu}$ and an isotopy f_t of the ns $|\underline{\xi}\times I|$ such that $f_1 : \underline{\mu}' \to \underline{\eta}\times I$ is a subdivision, and f_t is the identity on $|\underline{\xi}\times 0|$. Now we apply the same reasoning to the stratifications $\underline{\mu}'$ and the subdivision $(1\underline{\mu}'\times I)/(1\underline{M}'\times I)$ of $\underline{\zeta}\times I$, where $1\underline{\mu}'$ is defined to be $\underline{\mu}'\langle\underline{\xi}\times 1\rangle$. Thus there are a subdivision $\underline{\mu}''/\underline{M}''$ of $\underline{\mu}'$ and an isotopy g_t of the ns $|\underline{\xi}\times I|$ such that $g_1 : \underline{\mu}'' \to 1\underline{\mu}'\times I$ is a subdivision, and g_t is the identity on $|\underline{\xi}\times 1|$. Since the composites $f_1 : \underline{\mu}'' \to \underline{\mu}' \to \underline{\eta}\times I$ and $g_1 : \underline{\mu}'' \to 1\underline{\mu}'\times I \to \underline{\zeta}\times I$ are subdivisions, this proves the lemma.

Let X be a fixed polyhedron, and consider pairs (Y, X) with $Y \supset X$. We define (Y, X) and (Z, X) to be <u>equivalent</u> if there are regular neighbourhoods V of X in Y, W of X in Z, and a p.ℓ. isomorphism $f : V \to W$ which is the identity on X. More generally, let $\{\mathfrak{X}_n, \ldots, \mathfrak{X}_0\}$ be a variety of X. Then (Y, X) is a pair <u>over</u> $\{\mathfrak{X}_i\}$, and (Y, X) and (Z, X) are equivalent <u>over</u> $\{\mathfrak{X}_i\}$, if in addition, $\{\mathfrak{X}_i\}$ is a variety of X in Y and in Z. Thus if (Y, X) and (Z, X) are equivalent, they are equivalent over any variety of X in Y.

Let $\underline{\xi}/\underline{K}$ and $\underline{\eta}/\underline{L}$ be (n+1, n)-stratifications such that

$(|K_{n,}|, \ldots, |K_{o,}|) = (|L_{n,}|, \ldots, |L_{o,}|)$. Then $\underline{\xi}$ and $\underline{\eta}$ are <u>equivalent</u> if there are subdivisions $\underline{\xi}''/\underline{K}''$ of $\underline{\xi}$, $\underline{\eta}''/\underline{L}''$ of $\underline{\eta}$, and an isotopy f_t of $(|\underline{\xi}|, |K_{n,}|, \ldots, |K_{o,}|)$ such that $f_1 : \underline{K}'' \approx \underline{L}''$, and $f_1(\underline{\xi}'')$ is isomorphic to $\underline{\eta}''$ modulo $\underline{L}''$.

We can sum up the existence and uniqueness results of Chapters 2 and 3 as:

<u>Theorem 3.9.</u> Given a variety $\{\mathfrak{X}\}$ on a polyhedron X. Then there are bijections between:

the set $\mathcal{A}$ of equivalence classes of pairs (Y, X) over $\{\mathfrak{X}\}$;

the set $\mathcal{B}$ of equivalence classes of (n+1, n) stratifications over n-stratifications on $\{\mathfrak{X}\}$;

and the set $\mathcal{D}$ of isomorphism classes modulo $\underline{K}$ of (n+1, n)-stratifications over any fixed n-stratification $\underline{K}$ on $\{\mathfrak{X}\}$.

The bijection $a : \mathcal{A} \to \mathcal{B}$ is defined on representatives by:

$(Y, X) \to$ a stratification of $\{\mathfrak{X}\}$ in Y;

and $a : \mathcal{B} \to \mathcal{A}$ by: $\underline{\xi}/\underline{K} \to (|\underline{\xi}|, X)$.

The bijection $b : \mathcal{D} \to \mathcal{B}$ is defined by: the isomorphism class of $\underline{\xi} \to$ the equivalence class of $\underline{\xi}$.

The Corollary to Proposition 3.4 can be expressed as:

<u>Theorem 3.10.</u> Given polyhedra $Z \subset X \subset Y$ and a variety $\{\mathfrak{X}\}$ of X in Y, such that $Z < \mathrm{bdy}(X)$ (using the associated filtration). Let $h : (X) \to (Z) \times I$ be a fixed p.ℓ. isomorphism; thus we identify (X) with $(Z) \times I$. Then $Z < \mathrm{bdy}_Y(X)$; and if W is a neighbourhood of Z in bdy(Y, X), then there

is an equivalence $(Y, X) \sim (W, Z) \times I$ over (X), in which $(W, Z) \sim (W, Z) \times 0$ in the obvious way.

Corollary. Given X and a variety $\{\chi\}$ of X. Then whenever we have a pair $(Y, X \times D)$ over $\{\chi\} \times D$ (where D is a disk), there is some pair (Y', X) over $\{\chi\}$ such that $(Y, X \times D) \sim (Y', X) \times D$ over $\{\chi\} \times D$.

This theorem can be proved using Akin's results on homogeneous collapsing [1]. See also Lickorish and Siebenmann [13] for the case that Y is a manifold and $\{\chi\}$ the intrinsic variety of X. In Chapter 5 we shall generalize the Corollary, replacing the trivial disk block bundle $X \times D$ over X by any disk block bundle over X. (See also Chapter 8.)

Chapter 4. Transversality

Given polyhedra $X, Y \subseteq M$ with M a manifold, a variety $\{\mathfrak{X}\}$ of X in M, and a $(p+1, p)$-stratification $\underline{\xi}/\underline{K}$ of $\{\mathfrak{X}\}$ in M. Then Y is block transverse to $\underline{\xi}$ if

$Y \cap |\underline{\xi}| = \cup\{|\xi(\sigma)| : \text{some cells } \sigma \in \underline{K}''\}$.

We write: $Y \perp (X, \underline{\xi})$, or simply: $Y \perp \underline{\xi}$. Note that if $Y \perp \underline{\xi}$ and $\underline{\xi}'$ is a subdivision of $\underline{\xi}$, then $Y \perp \underline{\xi}'$.

Y is block transverse to X with respect to $\{\mathfrak{X}\}$ if there is some stratification $\underline{\xi}/\underline{K}$ of $\{\mathfrak{X}\}$ in M such that $Y \perp (X, \underline{\xi})$. We write: $Y \perp (X, \{\mathfrak{X}\})$, or simply: $Y \perp \{\mathfrak{X}\}$.

For most purposes one naturally takes $\{\mathfrak{X}\}$ to be the intrinsic variety of X. In this case we shall write: $Y \perp X$ if Y is block transverse to X with respect to its intrinsic variety. However, the proof of the transversality theorem does not depend on the variety of X used. Nor does the "local geometry" in the sense of the following discussion:

Let $Y \perp \underline{\xi}$. Consider $x \in \mathfrak{X}_i \cap Y$. Then we may express $st(x; \mathfrak{X}_i, \mathfrak{X}_i \cap Y) \underset{p.\ell.}{\approx} D \times c(E, G)$, with $x \longleftrightarrow v \times c$, where D is a disk, $v \in D$, (E, G) is not a cone- or suspension-pair, and E is either a disk or a sphere (since $\mathfrak{X}_i$ is an open manifold). If we take a $(p+1, p)$ stratification $\underline{\xi}^\wedge/\underline{K}^\wedge$ of $\{\mathfrak{X}\}$ in M with δ-small blocks, then $x \in |K^\wedge_{i,i}|$. Say $x \in \sigma$, for some $\sigma \in K^\wedge_{i,i}$, and let the fibre of $(\xi^\wedge, K^\wedge_{,i})(\sigma)$ be $c'(E', F')$, where E' is again a disk or a sphere, since M is a manifold. We express $c'(E', F') \underset{p.\ell.}{\approx} D' \times c'(E'', F'')$ where (E'', F'') is not a cone- or suspension-pair. By the Corollary to Lemma 3.2 we can choose $\underline{\xi}^\wedge/\underline{K}^\wedge$ so that $st(x; \mathfrak{X}_i, \mathfrak{X}_i \cap Y)$ is covered by a sub-complex of $K^\wedge_{i,i}$. By the local triviality of flags (see the Appendix, section I, 1), $(\xi^\wedge, K^\wedge_{,i})$ is trivial over $st(x; \mathfrak{X}_i, \mathfrak{X}_i \cap Y)$. Hence x has a neighbourhood in the quadruple $(M; X, Y; X \cap Y)$ of the form $D'' \times (cExc'E''; cExc'F'', cGxc'E''; cG \times c'F'')$, where $D'' = D \times D'$, and $x \longleftrightarrow v'' \times c \times c'$ for some $v'' \in D''$. (By the uniqueness of stratifications, $c'(E', F')$ is the fibre of $(\xi, K_{,i})$ over that component of $K_{i,i}$ which is in the same component of $\mathfrak{X}_i$ as x.)

This is essentially the condition that Y be polyhedrally transverse to X in M at x, as defined by Armstrong and Zeeman [4]. The difference is that they require $v'' \in \operatorname{int} D''$. The effect of our generalization is to allow transverse intersections between X and Y in which either (or both) may lie (locally) in bdy M. One can recover their situation by using a variety of X that refines its skeletal variety in M (our varieties always refine the intrinsic variety). Whichever of these situations we choose, we see that, locally at least, block transversality does not depend on what variety of X in M we use.

Remark. It is worth extracting from the foregoing discussion this observation: that if $\underline{\xi}/\underline{K}$ is an $(n+1, n)$-stratification, then $|\xi_{n+1,}|$ is a manifold if and only if $|\xi(\sigma)|$ is a disk for every cell σ of $\underline{K}$. $\xi(\sigma)$ may have fibre cS or cD.

A definition of transversality slightly stronger than block transversality will be very useful: Given $X, Y \subset M$ a manifold, a variety $\{\mathfrak{X}\}$ of

X in M and a (p+1, p)-stratification $\underline{\xi}$ of $\{X\}$ in M. Let $(Y_q, \ldots, Y_0)$ be a variety filtration of Y in M. Then (Y) is strongly transverse to $\underline{\xi}$ if each $Y_j \perp \underline{\xi}$, for $j = 0, \ldots, q$, and if $(Y \cap |\xi_{p+1,}|, \{Y_j \cap |\xi_{p+1,}|\})$ is a regular neighbourhood of $Y \cap X$ in (Y). We write: $(Y) \pitchfork \underline{\xi}$.

The second condition, though useful, is quite weak. Given a stratification $\underline{\xi}'/\underline{K}'$ of $\{X\}$ in M with $Y_j \perp \underline{\xi}'$ for all j, then we can "peel the frontier" off $\underline{\xi}'$ to obtain a stratification $\underline{\xi}/\underline{K}'$ with $(Y) \pitchfork \underline{\xi}$: Let $(V, \{V(\sigma) : \sigma$ is a cell of $\underline{K}\})$ be a regular neighbourhood (in fact a collar neighbourhood) of fr ξ' in $(|\xi'_{p+1,}|, \{|\xi'(\sigma)|\})$. Define a stratification $\underline{\xi}/\underline{K}$ by: $|\xi(\sigma)| = \mathrm{cl}[|\xi'(\sigma) - V(\sigma)|]$, $\partial\xi(\sigma) = |\xi(\sigma)| \cap \partial\xi'(\sigma)$, $\dot{\xi}(\sigma) = \cup\{|\xi(\rho)| : \rho \in \Lambda^*(\sigma, \underline{K})\} \cup (|\xi(\sigma)| \cap V(\sigma))$. We leave the reader to check that $\underline{\xi}/\underline{K}$ is indeed a stratification of $\{X\}$ in M. Now fr ξ is locally bicollared and two-sided in $(|\xi'_{p+1,}|, \{|\xi'(\sigma)|\}$, and hence also in (Y). By Cohen's Stellar Neighbourhood Theorem (quoted as our Theorem 1.3), $\underline{\xi}$ satisfies the desired property.

The proof of the transversality theorem requires a long digression on "decompositions" and "block decompositions".

Let K be a cell complex, p and q non-negative integers. A (p, q)-decomposition over K is a set of blocks $((\eta)) = \{\eta_{i,j}(\sigma) : \sigma \in K,\ i = 0, \ldots, p,\ j = 0, \ldots, q\}$ such that:

decomp 1 each $\eta_{i,j}$ is a cone block bundle over K;

decomp 2 for every $\sigma \in K$ there are compact polyhedra

$\emptyset = F_o \subseteq F_1 \subseteq \ldots \subseteq F_p$ and $\emptyset = G_o \subseteq G_1 \subseteq \ldots \subseteq G_q$, and a block structure $h : \eta_{p,q}(\sigma) \to \sigma \times cF_p \times cG_q$, which we identify with $\sigma \times c(F_p * G_q)$, so that h restricts to a block structure $\eta_{i,j}(\sigma) \to \sigma \times c(F_i * G_j)$ for every pair i, j.

<u>decomp 3</u> for every $\sigma \in K$, the filtrations $(F_p, \ldots, F_o)$ and $(G_q, \ldots, G_o)$ are variety filtrations.

For the proof of the transversality theorem we shall only need the case of (1, q)-decomposition. However, the discussion is no easier in this special case, so we shall continue in the general setting.

Observe that $((\eta))$ contains two distinguished flags over K: $(\xi) = (\eta_{p,o}, \eta_{p-1,o}, \ldots, \eta_{1,o})/K$ and $(\zeta) = (\eta_{o,q}, \eta_{o,q-1}, \ldots, \eta_{o,1})/K$ (recall our convention that if $(\xi_n, \ldots, \xi_1)/K$ is a flag, then $\xi_o = K$). $((\eta))$ will be called a decomposition of (ξ) and (ζ). We write: $((\eta))/K = ((\xi);(\zeta))/K$. However, (as we shall see in Chapter 6) $((\eta))$ is in no sense uniquely determined by (ξ) and (ζ).

We leave the reader to define an <u>isomorphism</u> of (p, q)-decompositions and a <u>trivial</u> decomposition.

An <u>admissable</u> isotopy of $((\eta))$ is a p.ℓ. isotopy h_t of $(|\eta_{p,q}|, \{|\eta_{i,j}|, \dot{\eta}_{p,q}\})$ rel $|K|$. h_t is <u>allowable</u> if h_t is the identity on all of $|(\xi)| \cup |(\zeta)|$. It follows from the Appendix, Theorem I, 4.A that given (p, q)-decompositions $((\eta))/K$, $((\eta'))/K'$ with $|\eta_{i,j}| = |\eta'_{i,j}|$ for all i, j (so in particular, $|K| = |K'|$), and <u>if</u> $|K|$ is a manifold, then there is an admissable isotopy h_t of $((|\eta|))$ rel K which carries some subdivision of $((\eta))$ isomorphically onto some subdivision of $((\eta'))$.

Further if either or both of the factors (ξ) and (ζ) of $((\eta))$ equal those of $((\eta'))$, then we may require h_t to be the identity there.

We shall henceforth assume of every decomposition $((\eta))/K$ that $|K|$ is a manifold.

An important class of decompositions is obtained thus: Let a p-flag $(\xi)/K$ and a q-flag $(\zeta)/K$ be given. Let Q be a cell complex triangulation of $(|\xi|)$ in which every block of (ξ) and its rim are covered by subcomplexes of Q. Then there exists a q-flag $(\nu)/Q$ such that $(\nu)\upharpoonright Q\langle K\rangle$ subdivides (ζ) (see the Appendix, Proposition II, 4.6 and Theorem II, 4.7). For every $\sigma \in K$ and each $i = 0, \ldots, p$, $j = 0, \ldots, q$, define the block $\eta_{i,j}(\sigma)$ by:

$$|\eta_{i,j}(\sigma)| = |\nu_j \upharpoonright Q\langle\xi_i(\sigma)\rangle|;$$

$$\partial\eta_{i,j}(\sigma) = |\nu_j \upharpoonright Q\langle\xi_i(\partial\sigma)\rangle|;$$

$$\dot{\eta}_{i,j} = \dot{\nu}_j \upharpoonright Q\langle\xi_i(\sigma)\rangle \cup |\nu_j \upharpoonright Q\langle\dot{\xi}_i(\sigma)\rangle|.$$

Then $\eta_{i,j}(\sigma)$ is indeed a block, and $((\eta)) = \{\eta_{i,j}(\sigma) : \sigma \in K, i = 0, \ldots, p, j = 0, \ldots, q\}$ is a (p,q)-decomposition over K, called a <u>block decomposition</u> of (ξ) and (ζ), and written: $((\eta)) = [(\xi); (\zeta)]/K$. $(\nu)/Q$ is called a <u>blocking</u> of $[(\xi); (\zeta)]$. Note that the definition of $[(\xi); (\zeta)]$ is apparently not symmetric in (ξ) and (ζ).

We shall also denote $((\eta))$ by: $[(\xi); (\nu)\upharpoonright Q\langle K\rangle]/K$. More generally, given flags $(\xi)/K$ and $(\zeta)/L$ with L a subdivision of K, let $(\zeta')/K$ be the amalgamation of (ζ). Then we shall often write $[(\xi); (\zeta)]/K$ for $[(\xi); (\zeta')]/K$. A similar convention will be used if K is a subdivision of L.

We show in the Appendix (Theorem II, 4.5) that given (ξ) and (ζ), $[(\xi);(\zeta)]$ is unique up to isomorphism of decompositions modulo $(\xi) \cup (\zeta)$; that is, it does not depend on the choice of Q and $(\nu)/Q$. We also show (Theorem II, 4.8) that there is one particular way to choose an isomorphism of decompositions which is natural:
$h : [(\zeta);(\xi)] \to [(\xi);(\zeta)]$ modulo $(\xi) \cup (\zeta)$; thus the definition of a block decomposition proves to be symmetric. This natural h is unique up to an allowable isotopy modulo $(\xi) \cup (\zeta)$ which respects the blocks of $((\eta))$ (Theorem I, 4.A); here again we need to know that $|K|$ is a manifold. We shall henceforth assume that $[(\xi);(\zeta)]$ and $[(\zeta);(\xi)]$ are identified by some such h.

Let $(\nu)/Q$ be a blocking of $((\eta)) = [(\xi);(\zeta)]/K$. Then we say an admissable (or allowable) isotopy f_t of $((|\eta|))$ <u>respects</u> $(\nu^{\bullet})$ if f_t restricts to an isotopy of $(\nu^{\bullet})$ (and hence f_t also restricts to an isotopy of $|(\nu) \restriction Q<(\xi^{\bullet})>|$). We shall also call f_t an <u>admissable</u> (or <u>allowable</u>) isotopy of (ν).

Let $((\eta)) = [(\xi);(\zeta)]/K$ be a block decomposition. Blockings $(\nu)/Q$ of $[(\xi);(\zeta)]$ and $(\mu)/P$ of $[(\zeta);(\xi)]$ are <u>complementary</u> if, for every $\sigma \in K$, there exist: trivializations of flags
$\mu h : (\mu) \restriction P<\sigma> \to P<\sigma> \times c(F)$ and $\nu h : (\nu) \restriction Q<\sigma> \to Q<\sigma> \times c(G)$;
a structure for $((\eta))(\sigma)$ $h : ((\eta))(\sigma) \to \sigma \times c(F) \times c(G)$ which extends μh and νh;
and allowable isotopies f_t, g_t of $((\eta))(\sigma)$,

such that:

$T \circ h \circ f_1 : (\mu) \restriction P\langle(\zeta)(\sigma)\rangle \to \sigma \times c(G) \times c(F)$

and $h \circ g_1 : (\nu) \restriction Q\langle(\xi)(\sigma)\rangle \to \sigma \times c(F) \times c(G)$

are trivializations of flags, where

$T : \sigma \times c(G) \times c(F) \to \sigma \times c(F) \times c(G)$ exchanges the factors.

Given μh and νh as above, then there always exists a structure h which extends μh and νh; and h is then unique up to allowable isotopy of $((\eta))(\sigma)$. If you choose different trivializations $\mu' h$ and $\nu' h$, then there is an isomorphism k of the trivial decomposition $\sigma \times c(F) \times c(G)$ which extends $\mu h \circ (\mu' h)^{-1}$ and $\nu h \circ (\nu' h)^{-1}$; and k is unique up to allowable isotopy. Also the identification of $[(\xi);(\zeta)]$ with $[(\zeta);(\xi)]$ is unique up to allowable isotopy of $((\eta))$. It follows that if $(\mu)/P$ and $('\mu)/P$ are both blockings of $((\eta))$ complementary to $(\nu)/Q$, and if $(\mu) \restriction P\langle K\rangle = ('\mu) \restriction P\langle K\rangle$, then there is an allowable isotopy a_t of $((\eta))$ such that $a_1 : (\mu) \approx ('\mu)$, and a_t respects the blocks of $((\eta))$. Conversely, if a_t is any such isotopy, then $a_1(\mu)$ is complementary to (ν). Another elementary property of complementary blockings is this: if (μ) and (ν) are complementary blockings of $((\eta))$, and $(\mu')/P'$, $(\nu')/Q'$ are subdivisions of (μ) and (ν), then (μ') and (ν') are complementary.

Given $((\eta))$ and (ν), then a complementary blocking (μ) always exists. In fact we can choose (μ) so that $(\nu^{\cdot}) = |(\mu) \restriction P\langle(\zeta^{\cdot})\rangle|$, and hence $|(\nu) \restriction Q\langle(\xi^{\cdot})\rangle| = (\mu^{\cdot})$. Then h, f_t, g_t, k and a_t of the previous discussion can always be chosen to respect $(\mu^\bullet) | P<(\eta)(\sigma)>$ and $(\nu^\bullet) | Q<(\xi)(\sigma)>$ for every $\sigma \in K$. (This follows from the uniqueness theorem for regular neighbourhoods of $(\xi^\bullet)(\sigma)$ in $((\eta))(\sigma)$.) Henceforth we shall usually assume that (μ) has been chosen in this way, and our isotopies - admissable and allowable - will usually respect $(\mu^\bullet)$ and $(\nu^\bullet)$.

This discussion of block decompositions is enough to prove the transversality theorem. If one tries to analyse block decompositions further, problems seem to arise about the nature of allowable isotopies which I cannot answer. These problems have to be solved, I believe, in order to prove block transversality symmetric. In Chapter 7 I shall try to elucidate some problems in the theory of block decompositions (See Chapter 8 for a further discussion of block decompositions, complementary blockings and transversality.)

Theorem 4.1. Given polyhedra $X, Y \subseteq M$ with M a manifold, a variety $\{\mathfrak{X}\}$ of X in M, and a $(p+1, p)$-stratification $\underline{\xi}/K$ of $\{\mathfrak{X}\}$ in M. Assume that $X \cap \text{bdy}\, M$ is $< \text{bdy}_M X$ and that $\underline{\xi}$ respects $X \cap \text{bdy}\, M$. Assume also that $Y \cap \text{bdy}\, M$ is $< \text{bdy}_M Y$. Then there are a subdivision $\underline{\xi}'/\underline{K}'$ of $\underline{\xi}$ and an isotopy f_t of M such that $f_1 Y \pitchfork \underline{\xi}'$.

Addendum 1. Let $(Y_q, \ldots, Y_o)$ be a variety filtration of Y in M such that $(Y) \cap \text{bdy}\, M < \text{bdy}(Y)$. Assume that for some $s = 0, \ldots, q$, $(Y_s) \pitchfork \underline{\xi}$; then we may add to the conclusion of the theorem: f_t is the identity on Y_s for all t.

Addendum 2. In the context of Addendum 1, $Y \cap \text{bdy}\, M$ inherits a variety filtration. Assume that $(Y \cap \text{bdy}\, M) \pitchfork \underline{\xi}$; then we may add to the conclusion of Addendum 1: f_t is the identity on bdy M for all t.

We shall concentrate on proving Addendum 2. Addendum 1 then follows by applying Addendum 2 first in bdy M (and extending the resulting isotopy of bdy M over M) and then in M. The main theorem follows from Addendum 1 by using the variety filtration (intrinsic filtration, $\emptyset$) of Y in M and setting $s = 0$.

Remark. Addendum 1 is trivially false if one requires only: $Y_s \pitchfork \underline{\xi}$. Take $M = \mathbb{R}^2$, $X = \mathbb{R}^1$, $(Y) = (Y, /, \cdot)$; then

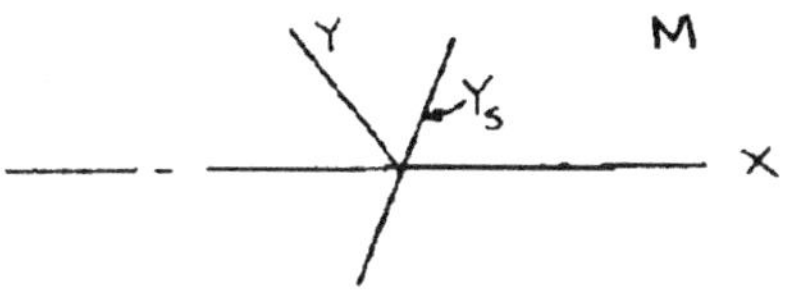

is a counter-example.

The proof starts with a number of reductions. Our goal is to reach the case in which X is a submanifold of M and Y is replaced, roughly speaking, by a neighbourhood in Y of an open manifold of the variety $\{\mathcal{Y}\}$ associated to the given filtration (Y). The reductions are not complete. In dealing with the special case we shall have to look back through the reductions from time to time. Some of these retrospects are incorporated in our analysis of the special case; others I have found more convenient to put after the special case is dealt with in its own right.

For each $i = 0, \ldots, p$, $\underline{\xi}$ includes a stratification of X_i in M (where (X) is the associated filtration to $\{\mathfrak{X}\}$ and we use the obvious variety of X_i), say $i\underline{\xi}/i\underline{K}$: $i\underline{\xi} = \cup\{\xi_{j,i'} : 0 \leq i' \leq i$ and $i' \leq j \leq i$ or $j = p+1\}$

Note that $(Y) \perp \underline{\xi} \Leftrightarrow (Y) \perp i\underline{\xi}$ for all i. Also if $(Y) \pm \underline{\xi}'$, a subdivision of $\underline{\xi}$, then $(Y) \pm$ the subdivision $i\underline{\xi}'$ of $i\underline{\xi}$ for all i. However, not every subdivision of $i\underline{\xi}$ comes from a subdivision of $\underline{\xi}$, so we shall have to choose our subdivisions carefully. This is one of the reasons to which I have alluded why we shall have to look backwards during our discussion of the special case. Let $\hat{i}$ be the smallest value of i for which (Y) is not strongly transverse to $i\underline{\xi}$; that is, $(Y) \pm (\hat{i}-1)\underline{\xi}$, but this is not true of $\hat{i}\underline{\xi}$. We shall make (Y) strongly transverse to $\hat{i}\underline{\xi}$ then by induction on $\hat{i}$ we shall be done. The advantage of this induction is that, by definition of block transversality, we have only to make Y transverse to the block bundle $\xi_{p+1,\hat{i}}/K_{\hat{i},\hat{i}}$; thus we may assume X is a manifold. The next reduction will make this statement precise:

Set $M^\wedge = \mathrm{cl}[M - |(\hat{i}-1)\xi_{p+1,}|]$, $X^\wedge = M^\wedge \cap X_{\hat{i}}$. Then $X^\wedge$ is a submanifold of $M^\wedge$, and has the normal disk block bundle $\xi_{p+1,\hat{i}}/K_{\hat{i},\hat{i}}$ in $M^\wedge$. Set $(Y^\wedge) = M^\wedge \cap (Y)$; then because of our strengthened notion of block transversality, $(Y^\wedge)$ is a variety filtration of $Y^\wedge$ in $M^\wedge$. Set $M^* = \mathrm{bdy}\, M^\wedge$, $X^* = \mathrm{bdy}\, X^\wedge$. Then $(M^*, X^*) < \mathrm{bdy}(M^\wedge, X^\wedge)$, and X^* has the normal disk block bundle

$\xi_{p+1,\hat{i}} \restriction K_{\hat{i},\hat{i}} \langle X^* \rangle$ in M^*, to which $Y^\wedge$ is strongly transverse by our inductive choice of i. Set $Y^\vee = Y^\wedge \cup M^*$, with variety $\{M^* - Y^\wedge, \psi^\wedge\}$. Then $M^* \cup Y^\wedge_s$ is strongly transverse to $\xi_{p+1,\hat{i}}$, by definition of s. By taking an equivalent filtration, we may think of $Y^{\vee}$ as having a variety filtration $(Y^{\vee}_{q+1}, \ldots, Y^{\vee}_{o})$ such that $(Y^{\vee}_{s+1})$ is already strongly transverse to $\xi_{p+1,\hat{i}}$. If we prove the theorem for $X^{\wedge}, Y^{\vee} \subseteq M^{\wedge}$, then the resulting isotopy $f^{\vee}_{t}$ of $M^{\wedge}$ keeps M^{*} fixed, and so extends to an isotopy f_t of M by the identity on $|(\hat{i}-1)\xi_{p+1,}|$. This completes our reduction to the case that X is a submanifold of M, and ξ/K a disk block bundle. We shall need to subdivide ξ/K. In practise the only conditions on these subdivisions will be on the cell complex K. By the Corollary to Lemma 3.2 we can always choose such subdivisions of ξ/K (which is really $\xi_{p+1,\hat{i}}/K_{\hat{i},\hat{i}}$) to come from subdivisions of $\underline{\xi}/\underline{K}$. Henceforth, unless otherwise stated, we shall assume this is always done. I shall usually denote a subdivision of ξ/K again by ξ/K.

Next we construct a convenient normal rns N of (Y) in M: Set $(Z') = (Y_s) \cap X$, and let $\underline{V}'$ be a normal $(s+1, s)$-rns of (Z') in X. Subdivide ξ/K so that each $V'_{j,i}$ and $V'^{\bullet}_{j,i}$ is covered by a subcomplex of K. Then set $V_{j,i} = |\xi \restriction K\langle V'_{j,i}\rangle|$, $V^{\bullet}_{j,i} = |\xi \restriction K\langle V'^{\bullet}_{j,i}\rangle|$, for $0 \leq i \leq s$, and $i \leq j \leq s$ or $j = q+1$. An application of Cohen's Stellar Neighbourhood Theorem (quoted as our Theorem 1.3) shows that $\underline{V} = \{V_{j,i}\}$ is a normal $(s+1, s)$-rns for $(Y_s) \cap |\xi|$ in $|\xi|$ with respect to $(\xi^{\bullet}, (Y') \cap \xi^{\bullet})$. Since $(Y_s) \cap \xi^{\bullet}$ is collared in $\mathrm{cl}[(Y') - |\xi|]$ - this is a typical use of our strengthened notion of block transversality - $\underline{V}$ extends to a normal $(s+1, s)$- rns $\underline{N}'$ of (Y_s) in M. Now $\underline{N}'$ extends to a normal $(q+1, q)$-rns $\underline{N}$ of (Y) in M. For let $\underline{N}^{\wedge}$ be any normal rns of (Y) in M; then there is an isotopy of (M, Y) carrying the normal $(s+1, s)$-rns

$\{N^{\wedge}_{j,i} : 0 \leq i \leq s,\ i \leq j \leq s \text{ or } j = q+1\}$ isomorphically onto $\underline{N}'$. Details are left to the reader; the plan is, first make $\underline{N}^{\wedge}$ agree with $\underline{N}'$ only as s-rns's for (Y_s) by an isotopy of (Y_s); then extend this isotopy to an isotopy of (M, Y); finally isotop M rel Y till $\underline{N}^{\wedge}$ agrees with $\underline{N}'$, using the uniqueness theorem for relative regular neighbourhoods.

Now let $\hat{j}$ be the smallest number such that either $(N_{q+1,\hat{j}}, \ldots, N_{\hat{j},\hat{j}})$ or $(N^{\bullet}_{q+1,\hat{j}}, \ldots, N^{\bullet}_{\hat{j},\hat{j}})$ is not strongly transverse to ξ. Henceforth I shall abbreviate $N_{r,\hat{j}}$ and $N^{\bullet}_{r,\hat{j}}$ to N_r and $N^{\bullet}_r$, for $r = \hat{j}, \ldots, q+1$. Set $N^*_r = (N_r \cap \text{bdy } M) \cup \{N^{\bullet}_{r,j} : 0 \leq j \leq \hat{j}-1\}$; then $N^*_r < \text{bdy } N_r$ with $N^*_{\hat{j}} = \text{bdy } N_{\hat{j}}$ and $(N^*) \pitchfork \xi$. We shall make (N) strongly transverse to ξ keeping (N^*) fixed; this will be the inductive step in an induction on $\hat{j}$, which suffices to prove the theorem.

Just as we "threw away" $(\hat{i}-1)\underline{\xi}$, we now throw away $\cup\{N_{q+1,j} : j \leq \hat{j}-1\}$; that is, replace M by $\text{cl}[M - \cup N_{q+1,j}]$ and so on. This reduces us to the case that (Y) is a family $(N_{q+1}, \ldots, N_{\hat{j}})$ such that $(N^*) \pitchfork \xi$, and $N_{\hat{j}}$ is a submanifold of M with $N^*_{\hat{j}} = \text{bdy } N_{\hat{j}} = N_{\hat{j}} \cap \text{bdy } M$. We have to make (N) strongly transverse to ξ, keeping bdy M fixed.

We have now finished all the reductions, and are entering upon the core of the proof. By Rourke and Sanderson's relative transversality theorem [19, II] for two manifolds in a manifold, there are a subdivision ξ'/K' of ξ and an isotopy a_t of M rel bdy M such that $a_1 N_{\hat{j}} \perp \xi'$. It is easily seen that we can make $a_1 N_{\hat{j}}$ transverse to ξ' in our stronger sense (see the proof of their theorem). I claim we can choose ξ' to come from a subdivision of $\underline{\xi}/\underline{K}$. For first we can choose $K'<\mathrm{bdy}\ M>$ to equal $K<\mathrm{bdy}\ M>$, since there we already have $(N^*) \perp \xi$, by using a collar of bdy M in M. Next, let ξ''/K' be a subdivision of ξ which does come from a subdivision $\underline{\xi}''$ of $\underline{\xi}$. Then $\xi'' \vdash K'<\mathrm{bdy}\ M> = \xi' \vdash K'<\mathrm{bdy}\ M>$. So there is an isotopy of M rel $|K| \cup$ bdy M carrying ξ' isomorphically to ξ''. Then $a_1 N_{\hat{j}}$ is carried strongly transverse to ξ'', as required. Henceforth I shall suppress a_t and ξ'', and assume $N_{\hat{j}}$ is already strongly transverse to ξ. Of course, if $\hat{j} \leq s$, then $N_{\hat{j}} \subseteq Y'$, and this step is unnecessary. During the rest of the proof we shall keep $N_{\hat{j}}$ fixed. To simplify notation, I shall denote $\max(s, \hat{j})$ simply by: s. This reduces us to the (more difficult) case that $\hat{j} \leq s$.

The next step uses the fact that block transversality is symmetric for two manifolds in a manifold (see [19, II] or the Appendix, Theorem II, 4.8). We shall construct a normal disk block bundle η/H of $N_{\hat{j}}$ in M such that $X \perp \eta$ and η accords with (N^*) and $(N_s, \ldots, N_{\hat{j}})$, which are already strongly transverse to ξ: Since $(N^*) \perp \xi$, $(N^*) \cap X$ is a variety filtration, as can be seen from our local p. ℓ. analysis of block transversality. Hence there is a flag structure $(''\eta^*) = (''\eta^*_{q+1}, \ldots, ''\eta^*_{\hat{j}+1})/''H^*$ on $(N^*) \cap X$. Let W be a regular neighbourhood of $N_s \cap X$ in X rel $\dot{N}_s \cap X$ which meets bdy X regularly in $|''\eta^*_{q+1}|$. Then $(W, N_s \cap X, \ldots, N_{\hat{j}} \cap X)$ is a variety filtration too, and so can be given the structure of a flag $(''\eta) = (''\eta_{q+1}, ''\eta_s, \ldots, ''\eta_{\hat{j}+1})/''H$. We may assume $(''\eta)$ extends the flag $(''\eta^*_{q+1}, ''\eta^*_s, \ldots, ''\eta^*_{\hat{j}+1})/''H^*$. Subdivide ξ so that every block of $(''\eta^*)$ and $(''\eta)$, and the rim of every such block, are covered by subcomplexes of K. Then $\xi \restriction K\langle''\eta^*_{q+1}\rangle$ defines a block decomposition $((\lambda^*)) = [(''\eta^*); \xi \restriction K\langle''H^*\rangle]/''H^*$. Let $('\eta^*)/'H^*$ be a complementary blocking. Similarly $\xi \restriction K\langle''\eta_{q+1}\rangle$ defines a block decomposition $((\lambda)) = [(''\eta); \xi \restriction K\langle''H\rangle]/''H$. Let $('\eta)/'H$ be a complementary blocking; we may assume it extends $('\eta^*_{q+1}, '\eta^*_s, \ldots, '\eta^*_{\hat{j}+1})/'H^*$. Since $|'H|$ is a regular neighbourhood of $N_{\hat{j}} \cap X$ in $N_{\hat{j}}$, it follows that $'\eta_{q+1}/'H$ extends to a normal disk block bundle η/H of $N_{\hat{j}}$ in M; this is the required block bundle.

We now choose a flag structure $(\nu) = (\nu_{q+1}, \ldots, \nu_{\hat{j}+1})/H$ on $(N_{q+1}, \ldots, N_{\hat{j}})$. By the uniqueness theorem for normal flags, we shall be able to isotop M till the sub-flag $(\nu_{q+1}, \nu_s, \ldots, \nu_{\hat{j}+1})/H$ equals $(\eta)/H$. The effect of this step, then, will be to improve from having $X \perp (N_{\hat{j}}, \eta)$ to having X (approximately) block transverse to the whole system $(N_{q+1}, \ldots, N_{\hat{j}})$: (ν) is chosen to extend $('\eta^*)/'H^*$ and such that $(\nu_s, \ldots, \nu_{\hat{j}+1})/H = (\eta_s, \ldots, \eta_{\hat{j}+1})/H$. This can be done, since any choice of (ν) can be isotoped till it meets these conditions (see the Appendix, Theorem

I, 4.A). There is now an isotopy c_t of M rel bdy M $\cup$ N_s such that $c_1 : (\nu_{q+1}, \nu_s, \ldots, \hat{\nu}_{j+1})/H \approx (\eta_{q+1}, \eta_s, \ldots, \hat{\eta}_{j+1})/H$, by the same theorem of the Appendix. Henceforth I shall suppress c_t and assume that $(|\xi|, X)$ is already "transverse" to the flag $(\nu)/H$. (I leave the reader to define "transversality to a flag".)

The final step starts with choosing a disk block bundle $''\xi/''K$ complementary to $(\nu) \upharpoonright 'H$. Then $(N_{q+1}) \pitchfork ''\xi$. There is a restriction ξ''/K'' of ξ such that $|''\xi| = |\xi''|$, namely: $K'' = K\langle\nu_{q+1} \upharpoonright ''H\rangle$. By subdividing ξ and $''\xi$ we may assume $K'' = ''K$. We isotop $''\xi$ till it equals ξ''. Then $(N_{q+1}) \pitchfork \xi$, as required: The flag $(\nu) \upharpoonright 'H$ defines a block decomposition $((\lambda^{\wedge})) = [(\nu) \upharpoonright 'H\langle''H\rangle; \xi \upharpoonright K\langle''H\rangle]/''H$, which extends $((\lambda^*))$ and contains $((\lambda))$ as a sub-family of block bundles. Let $''K$ be a subdivision of K in which every block of $(\nu) \upharpoonright 'H\langle''H\rangle$ and its rim are covered by subcomplexes. Let $\xi''/''K\langle''H\rangle$ be a subdivision of $\xi \upharpoonright K\langle''H\rangle$ such that $\xi + \xi''$ comes from a subdivision of $\underline{\xi}/\underline{K}$. Then ξ'' extends to a blocking $''\xi/''K$ of $((\lambda^{\wedge}))$ complementary to $(\nu) \upharpoonright 'H$. I claim we can choose $''\xi$ to agree, wherever (N) is already transverse to ξ, with the restriction of some subdivision of ξ that comes from a subdivision of $\underline{\xi}/\underline{K}$. First, let $''\xi^*/''K^*$ be the restriction of $''\xi$ to $''K\langle\nu_{q+1} \upharpoonright ''H^*\rangle$ in bdy M. Let $\xi^*/''K^*$ be a subdivision of ξ such that $\xi + \xi^*$ comes from a subdivision of $\underline{\xi}/\underline{K}$ and which extends $\xi'' \upharpoonright ''K\langle''H^*\rangle$. (The proof of the Corollary to Lemma 3.2 shows this can be done.) Then $''\xi^*$ and ξ^* are both blockings of $((\lambda^*))$ complementary to $(\nu) \upharpoonright 'H^*$, which agree in $|'H^*|$. So there

is an allowable isotopy of $((\lambda^*))$ which carries $''\xi^*$ isomorphically to ξ^*. Thus we may assume $''\xi^*$ already comes from (the restriction of) a subdivision of $\underline{\xi}/\underline{K}$. Second, let $s''\xi/s''K$ be the restriction of $''\xi$ to $''K\langle\nu_s \vdash ''H\rangle$. Let $s\xi/s''K$ be a subdivision of $\xi \vdash K\langle\nu_s \vdash ''H\rangle$ such that $\xi + s\xi$ comes from a subdivision of $\underline{\xi}/\underline{K}$. We may assume $s\xi$ extends $''\xi \vdash ''K\langle\nu_s \vdash ''H^*\rangle$ and $''\xi \vdash ''K\langle''H\rangle$. Then $s''\xi$ and $s\xi$ are both blockings of the block decomposition $((s\lambda)) = [(\nu_s, \ldots, \nu_{\hat{j}+1})/''H;\ ''\xi \vdash ''K\langle''H\rangle]/''H$ which are complementary to $(\nu_s, \ldots, \nu_{\hat{j}+1})/''H$. So there is an allowable isotopy h_t of $((s\lambda))$ such that $h_1 : s''\xi \approx s\xi$ and h_t is the identity on $|((s\lambda))| \cap \text{bdy}\, M$. Now for each cell $\sigma \in ''H$ we have the variety filtration of $|((\lambda^\wedge))(\sigma)| = |\lambda_{q+1,1}(\sigma)|$ by:
$(|\lambda_{q+1,1}(\sigma)|, |\lambda_{q,1}(\sigma)| \cup |\lambda_{q+1,0}(\sigma)|, \ldots, |\lambda_{\hat{j},1}(\sigma)| \cup |\lambda_{q+1,0}(\sigma)|,$ $|\lambda_{q+1,0}(\sigma)|, |\lambda_{q,0}(\sigma)|, \ldots, |\lambda_{\hat{j}+1,0}(\sigma)|, \sigma)$. Since h_t extends to an isotopy of the sub-filtration $(|\lambda_{s,1}(\sigma)| \cup |\lambda_{q+1,0}(\sigma)|)$ by the identity on $|\lambda_{q+1,0}(\sigma)|$, it follows that h_t extends to an isotopy k_t of the whole filtration which is the identity on $|\lambda_{q+1,0}(\sigma)|$. Since h_t is already the identity on $|\lambda_{\hat{j},1}(\sigma)|$, k_t is an allowable isotopy of $((\lambda^\wedge))(\sigma)$. Using induction on a standard ordering of $''H$ and a standard procedure, we can extend h_t to an allowable isotopy of $((\lambda^\wedge))$ which is the identity on $|((\lambda^\wedge))| \cap \text{bdy}\, M$. Thus we may assume that $s''\xi$ comes from (the restriction of) a subdivision of $\underline{\xi}/\underline{K}$. This justifies our claim, and completes the construction of $''\xi/''K$.

Now let $\xi^\wedge/''K$ be a subdivision of $\xi \vdash K\langle\nu_{q+1} \vdash ''H\rangle$ such that

$\xi + \xi^{\wedge}$ comes from a subdivision of $\underline{\xi}/\underline{K}$. We may assume that $\xi^{\wedge}$ extends $''\xi \vdash ''K<\nu_{q+1} \vdash ''H*>$ and $''\xi \vdash ''K<\nu_s \vdash ''H>$. Then we have $|\xi^{\wedge}| = |''\xi|$, and $\xi^{\wedge}$ and $''\xi$ are both blockings of $((\lambda\ )) =$ $= [(\nu_{q+1}, \nu_s, \ldots, \nu_{\hat{j}+1}) \vdash ''H\ ; \ \xi \vdash K<''H>]/''H$ which are complementary to $(\nu_{q+1}, \nu_s, \ldots, \nu_{\hat{j}+1}) \vdash 'H$. (This is true of $\xi^{\wedge}$ by construction of $('\eta)$, and of $''\xi$ by construction of $''\xi$.). So there is an allowable isotopy $d_t^{\wedge}$ of $((\lambda^{\wedge}))$ such that $d_1^{\wedge} : ''\xi \approx \xi^{\wedge}$. In fact we can choose $d_t^{\wedge}$ to be the identity on $|\nu_s \vdash 'H|$, since $\xi^{\wedge}$ already equals $''\xi$ there. The proof of this statement is similar to the way in which we made $s''\xi$ agree with $s\xi$. Details are left to the reader; the plan is to start with any allowable isotopy $d_t^{\wedge}$ of $((\lambda))$ such that $d_1^{\wedge} : ''\xi \approx \xi$. Consider $((\lambda)) \times I/''H \times I$; then $(\nu\ ) \times I$ is a blocking one way, and both $\xi^{\wedge} \times I$ and $D^{\wedge}(''\xi \times I)$ are complementary blockings, where $D^{\wedge}$ is the p. ℓ. isomorphism of $|((\lambda\ ))| \times I$ defined by $D^{\wedge}(x, t) = (d_t^{\wedge}(x), t)$. Find an allowable isotopy $D_t^{\wedge}$ of $((\lambda\ )) \times I$ rel $((\lambda\ )) \times 0$ and $((\lambda \times 1))$ so that $D_1^{\wedge}(''\xi \times I) = \xi^{\wedge} \times I$. $D_t^{\wedge}$ can be chosen to be an allowable isotopy of $((\lambda^{\wedge})) \times I$, by our previous argument using a variety filtration. And $D_t^{\wedge}$ can be chosen to be level-preserving. So $D_1^{\wedge}$ is the trace of an allowable isotopy of $((\lambda^{\wedge}))$, which is the required improved choice of $d_t^{\wedge}$.

We have only to extend $d_t^{\wedge}$ to an isotopy of M rel bdy M. It suffices to extend $d_t^{\wedge}$ to an isotopy of $(|\xi|, X)$ rel $|\xi| \cap$ bdy M. This can be done if $(|\xi|, X)$ and $(|\xi| \cap \text{bdy } M, \text{bdy } X)$ are strongly transverse to $(N_{\hat{j}}, \nu_{q+1})$ (and not just transverse, as we proved above). I shall prove the first of these conditions; the second is proved similarly. From the way we defined the normal rns $\underline{N}$ of Y in M in our original problem, from which rns (N_{q+1}) is derived, it follows that $X \pitchfork \nu_{q+1}$. From the construction of $('\eta)/'H$ we see that $|\nu_{q+1} \nmid 'H| = |\xi \nmid K\langle W\rangle|$, where $W = X \cap |\nu_{q+1}|$. Hence $(|\xi|, X) \pitchfork \nu_{q+1}$, as required. So $\hat{d}_t$ extends to an isotopy d_t of M rel bdy M. This completes the final step.

Composing the isotopies a_t, c_t and d_t gives the isotopy f_t needed to establish the inductive step.

a_t, c_t and d_t are all the identity on bdy M in our reduced problem. Tracing back through the reductions to the original problem, we see that our inductive step leaves bdy M fixed. This proves Addendum 2, in case $Y_s = \emptyset$.

In the general case, $Y_s \neq \emptyset$, we have to go back and check that a_t, c_t and d_t can be chosen to be the identity on Y_s (and not just $N_{s,\hat{j}}$). $Y_{\hat{j}-1}$ was "thrown away" and so is left fixed. Now if $\hat{j} \geq s$ there is nothing to prove. If $\hat{j} \leq s$, then the isotopy a_t is not needed, and may be taken to be the identity. c_t is an isotopy of $(N_{q+1,\hat{j}}, \ldots, N_{\hat{j},\hat{j}})$ rel $N_{s,\hat{j}}$. Extend c_t first over $Y_s \cup$ bdy M by the identity. Then c_t on $Y_s \cup \text{bdy } M \cup N_{q+1,\hat{j}}$ respects the variety filtration $(Y_s \cup \text{bdy } M \cup N_{q+1,\hat{j}}, Y_s \cup N_{q+1,\hat{j}}, Y_{s-1} \cup N_{q+1,\hat{j}}, \ldots, Y_{\hat{j}+1} \cup N_{q+1,\hat{j}}, N_{q+1,\hat{j}}, N_{s,\hat{j}}, N_{s-1,\hat{j}}, \ldots, N_{\hat{j},\hat{j}})$ of $Y_s \cup \text{bdy } M \cup N_{q+1,\hat{j}}$ in M. So c_t extends to an isotopy of M rel $Y_s \cup$ bdy M as required. d_t was defined first as an isotopy $\hat{d}_t$ of $|''\xi|$ rel $N_{s,\hat{j}} \cap |''\xi|$. We then observed that $(|''\xi|, |''\xi \nmid ''K\langle N_{s,\hat{j}} \cap X\rangle, \ldots, |''\xi \nmid ''K\langle N_{\hat{j},\hat{j}} \cap X\rangle|)$ is a regular neighbourhood of $Y_{\hat{j}} \cap |\xi|$ in $(|\xi|, Y_s \cap |\xi|, \ldots, Y_{\hat{j}} \cap |\xi|)$ which meets bdy $M \cap |\xi|$ regularly. A similar argument to that given for c_t shows we may extend $d_t^{\wedge}$ to an isotopy $d_t^{\vee}$ of $|\xi|$ rel $(Y_s \cap |\xi|)$ and rel (bdy $M \cap |\xi|$). Finally, our assumption that $(Y_s) \pitchfork \xi$ implies that $(|\xi|, Y_s \cap |\xi|, \ldots, Y_{\hat{j}} \cap |\xi|)$ is a regular neighbourhood of X in $(M, Y_s, \ldots, Y_{\hat{j}})$ which meets bdy M regularly. So the same argument shows that $d_t^{\vee}$ extends to an isotopy d_t of M rel $Y_s \cup$ bdy M, as required. This completes the proof of Addendum 2, and hence of Addendum 1 and Theorem 4.1.

Corollary. Given polyhedra X, $Y \subset M$ a manifold, and a variety $\{\mathfrak{X}\}$ of X in M. Assume that $X \cap$ bdy $M <$ $\text{bdy}_M X$ and $Y \cap$ bdy $M <$ $\text{bdy}_M Y$. Then there is an ε-isotopy f_t of M such that $f_1 Y$ is block transverse to $(X, \{\mathfrak{X}\})$, and we may assume f_t is the identity outside any given neighbourhood of X in M. If $Y \cap$ bdy M is already block transverse to $X \cap$ bdy M in bdy M, then we may assume f_t is the identity on bdy M.

The proof is left to the reader. One must choose a stratification $\xi/\underline{K}$ of $\{\mathfrak{X}\}$ in M with δ-small blocks such that $Y \cap$ bdy $M \perp \underline{\xi} \restriction \underline{K}\langle X \cap \text{bdy } M\rangle$, and then check that all our isotopies can be made small.

To phrase Addendum 1 in this context requires some care: one must specify that $Y_s, \ldots, Y_o$ are block transverse to the <u>same</u> stratification of $\{\mathfrak{X}\}$ in M. I leave the reader to state the appropriate Corollary.

Chapter 5. Classifying Stratifications and Cobordism Theories

In this chapter we define and discuss "stratifiable maps" between polyhedra. First we shall define a class of maps, called "stratified maps", between n-stratifications in such a way that we can "pull back" any (n+1, n)-stratification over the image to an (n+1, n)-stratification over the domain. This is in accordance with our intuition of an (n+1, n)-stratification $\underline{\eta}/\underline{L}$ as a "generalized bundle" over $\underline{L}$. Then we shall define a "stratifiable map" between polyhedra X and Y (roughly) as a p.ℓ. map $f: X \to Y$ which can be expressed as a stratified map between some choice of stratifications of X and Y. Any abstract regular neighbourhood M of Y can then be "pulled back" to an abstract regular neighbourhood f^*M of X. If M is a manifold, then so is f^*M.

Stratifiable maps are, in some sense, "isomorphisms of local geometry". As we shall see in Chapter 6, the class of stratifiable maps is not the most general to satisfy this property. Nor is it the most general class of maps such that one can pull back abstract regular neighbourhoods. And the class of stratified maps is not so general as Thom's class of "morphismes stratifies" [28]. One of the "open questions" I shall mention in Chapter 7 is to define more general maps between polyhedra than stratifiable maps (See also Chapter 8.)

There are four candidates for a "stratified morphism" between n-stratifications:

i. if $\underline{L}$ is an n-stratification, $\underline{K}$ a restriction of $\underline{L}$, then the inclusion of $\underline{K}$ in $\underline{L}$ should be a stratified map, since we can restrict any $(n+1, n)$-stratification $\underline{\eta}/\underline{L}$ to $\underline{K}$;

ii. if $\underline{L}'$ is a subdivision of $\underline{L}$, then the identity map of L should be a stratified map both from $\underline{L}'$ to $\underline{L}$ -

iii. - and from $\underline{L}$ to $\underline{L}'$, since we can subdivide any $\underline{\eta}/\underline{L}$ to some $\underline{\eta}'/\underline{L}'$, and amalgamate any $\underline{\eta}'/\underline{L}'$ to some $\underline{\eta}/\underline{L}$;

iv. the projection $\underline{L} \times I \to \underline{L}$ should be a stratified map, since any $\underline{\eta}/\underline{L}$ induces $\underline{\eta} \times I/\underline{L} \times I$.

The definition we use is a combination of all four types: If $\underline{\xi}$ and $\underline{\eta}$ are (q, n)-bs's, a <u>(q, n)-bs map</u> $f : \underline{\xi} \mapsto \underline{\eta}$ is a p.ℓ. map $f : |\xi_{q,}| \to |\eta_{q,}|$ such that there exist:

i. a cell complex triangulation B of a disk D^u;

ii. a subdivision $(\underline{\eta} \times B)''$ of the (q, n)-bs $\underline{\eta} \times B$;

iii. a sub-bs $\underline{\xi}'' < (\underline{\eta} \times B)''$;

iv. a subdivision $f'' : \underline{\xi}'' \to \underline{\xi}$;

such that $f = p \circ f''^{-1}$, where $p : |\eta_{q,} \times B| \to |\eta_{q,}|$ is the projection.

We shall refer to $(D^u, B, (\underline{\eta} \times B)'', \underline{\xi}'', f'')$ as a <u>structure</u> for f; we shall omit some of these terms from the list if they are clear from context.

(q, n)-bs maps $f_o : \underline{\xi} \mapsto \underline{\eta}$ and $f_1 : \underline{\xi} \mapsto \underline{\eta}$ are <u>concordant</u> if there is a (q, n)-bs map $F : \underline{\xi} \times I \mapsto \underline{\eta} \times I$ such that $F^{-1}(|\eta_{q,}| \times \{0, 1\}) = |\xi_{q,}| \times \{0, 1\}$, and F equals $f_o \times 0$ on $|\xi_{q,}| \times 0$ and $f_1 \times 1$ on

$|\xi_{q,}| \times 1$.

Note that if ξ and η are 0-bs's, that is, cell complexes, then any p.ℓ. map $f : |\xi| \to |\eta|$ is a 0-bs-map. For any p.ℓ. map between polyhedra $f : X \to Y$ can be expressed as the composition $p \circ g$, where $g : X \to Y \times D$ is an embedding, D being a large-dimensional disk, and where $p : Y \times D \to Y$ is the projection. Similarly any p.ℓ. homotopy of $|\xi|$ to $|\eta|$ is a 0-bs concordance.

Henceforth we shall assume (unless otherwise stated) that ξ and η are either n-stratifications or (n+1, n)-stratifications, and we shall call $f : \xi \mapsto \eta$ a <u>map of stratifications</u> or a <u>stratified map.</u>

<u>Lemma 5.1.</u> If $f : \xi \mapsto \eta$ and $g : \eta \mapsto \zeta$ are stratified maps, then f is concordant to an f_1 such that $g \circ f_1$ is a stratified map.

Let $(D^u, B, (\eta \times B)'', \xi'', f'')$ be a structure for f, and $(D^t, C, (\zeta \times C)', \eta', g')$ a structure for g. By Lemma 3.8 there exist a subdivision θ^* of $(\eta \times B) \times I + (\eta \times B)'' \times 0 + (\eta' \times B) \times 1$ and subdivisions $a : \theta^* \to (\eta \times B)'' \times I$, $b : \theta^* \to (\eta' \times B) \times I$, such that a is the identity on $|\eta_{q,}| \times |B| \times 0$, b the identity on $|\eta_{q,}| \times |B| \times 1$. Say $1\theta^*$ is the subdivision of $(\eta' \times B) \times 1$ induced by θ^*; let $(\xi \times I)^*$ be the subdivision $\theta^* \langle a^{-1}(|\xi''_{q,}| \times I) \rangle$ of $\xi'' \times I$; and set $1\xi^* = 1\theta^* \cap (\xi \times I)^*$. Let $p^* : |\eta_{q,}| \times |B| \times I \to |\eta_{q,}| \times I$ be the projection. Let F be the composite

$$|\xi_{q,}| \times I \xrightarrow[f''^{-1} \times id]{} |\xi''_{q,}| \times I \subseteq |(\eta \times B)''_{q,}| \times I \xrightarrow[a^{-1}]{} |\theta^*_{q,}| \xrightarrow[p^*]{} |\eta_{q,}| \times I.$$

Then F is a concordance of f to some f_1, for F has the structure $(B \times I, \theta^*, (\xi \times I)^*, F^* = (f'' \times id.) \circ a)$. Set $f_1^* = F^* \restriction |1\xi^*_{q,}|$. Then

$(1\underline{\theta}^*, 1\underline{\xi}^*, f_1^*)$ is a structure for f_1. Note that $1\underline{\theta}^*$ is a subdivision of $\underline{\eta}' \times B \times 1$, since b is the identity there. So identifying $\underline{\eta}' \times B \times 1$ with $\underline{\eta}' \times B < (\underline{\xi} \times C)' \times B$, $(D^{u+t}, B\times C, (\underline{\xi} \times C)'\times B + 1\underline{\theta}^*, 1\underline{\xi}^*, f_1^* \circ (g' \times id_{D^u}))$ is a structure for $g \circ f_1$, which is thus a stratified map.

Lemma 5.2. Given $f : \underline{\xi} \mapsto \underline{\eta}$. Let $(D^u, B, (\underline{\eta} \times B)'', \underline{\xi}'', f'')$ and $(D^u, B, (\underline{\eta} \times B)', \underline{\xi}', f')$ be structures for f. Then $g'' = f''^{-1} : \underline{\xi} \mapsto \underline{\eta} \times B$ and $g' = f'^{-1} : \underline{\xi} \mapsto \underline{\eta} \times B$ (which clearly are stratified maps) are concordant.

Let A be a cell complex triangulation of a disk D^t, where t is large, say $t \geq 2(\dim |\underline{\xi}_{q,}| + 1)$; thus $C = B \times A$ triangulates D^{u+t}. Let $\underline{\mu}$ be obtained from $\underline{\eta} \times C \times I$ by subdividing $\underline{\eta} \times C \times 0$ to $\underline{\mu}'' = (\underline{\eta} \times B)'' \times A \times 0$ and $\underline{\eta} \times C \times 1$ to $(\underline{\eta} \times B)' \times A \times 1$. Assume inductively that for each $j \leq i-1$ we have found a subdivision $j\underline{\mu}$ of $\Lambda(\mu_{j,j}, \underline{\mu})$ (which is a weak $(n-j+1, n-j)$-stratification, a restriction $j\underline{\nu}$ of $j\underline{\mu}$ and a subdivision $jh : j\underline{\nu} \to \Lambda(\xi_{j,j} \times I, \underline{\xi} \times I)$, such that, whenever $k < j$:

$j\underline{\mu}$ restricts to a subdivision of $\Lambda(\Lambda_k^*(k\underline{\mu})_{j,j}, k\underline{\mu})$

$j\underline{\nu}$ restricts to a subdivision of $\Lambda(\Lambda_k^*(k\underline{\nu})_{j,j}, k\underline{\mu})$, and jh extends kh there;

$j\underline{\nu}$ meets $\underline{\mu}''$ and $\underline{\mu}'$ in subdivisions of $\Lambda(\mu''_{j,j}, \underline{\mu}'')$ and $\Lambda(\mu'_{j,j}, \underline{\mu}')$.

Thus $\underline{\mu}$ has the subdivision $\underline{\mu}^\wedge = \underline{\mu} + 0\underline{\mu} + \ldots + (i-1)\underline{\mu}$. If $\underline{\lambda}$ is an $(n+1, n)$-stratification (such as $\underline{\mu}^\wedge$, $\underline{\xi} \times I$, etc.), define

$\partial\lambda_{i,i} = \cup\{\Lambda^*(\lambda_{j,j}, \underline{\lambda})_{i,i} : j \leq i-1\}$; thus $\partial\lambda_{i,i}$ is a cell complex $<$ bdy $\lambda_{i,i}$. Similarly we define $\partial\nu_{i,i} = \cup\{\Lambda^*(\nu_{j,j}, \underline{\mu}^\wedge)_{i,i} : j \leq j-1\}$ Then the jh fit together to define a subdivision

$\partial h : \Lambda(\partial \nu_{i,i}, \underline{u}^{\wedge}) \to (\partial(\xi_{i,i} \times I)$.

Now $|u_{i,i}| = |\eta_{i,i}| \times D^{u+t} \times I$; let $p : |u_{i,i}| \to |\eta_{i,i}|$ be the projection. Then the maps $p\circ(f''^{-1} \times 0 \vdash)$, $p\circ(f'^{-1} \times 1 \vdash) : |\xi_{i,i}| \to |\eta_{i,i}|$ both equal $f : |\xi_{i,i}| \to |\eta_{i,i}|$. So $f''^{-1} \times 0 \vdash$ and $f'^{-1} \times 1 \vdash :$ $: |\xi_{j,i}| \to |u_{i,i}|$ are homotopic. Since t is large, there is an embedding $a : |\xi_{i,i}| \times I \hookrightarrow |u_{i,i}|$ which extends $f''^{-1} \times 0 \vdash$ and $f'^{-1} \times 1 \vdash$. We may also assume that a extends $\partial h^{-1} : |\partial(\xi_{i,i} \times I)| \to |\partial \nu_{i,i}| \subseteq |\partial u_{i,i}|$. Take a cell complex subdivision $u^{*}_{i,i}$ of $u^{\wedge}_{i,i}$ such that $a(|\xi_{i,i}| \times I)$ is covered by a subcomplex $\nu^{*}_{i,i} < u^{*}_{i,i}$ and $a^{-1} : \nu^{*}_{i,i} \to \xi_{i,i} \times I$ is a subdivision. Let $i\underline{u}$ be a subdivision of $\underline{u}^{\wedge}$ such that $iu_{i,i} = u^{*}_{i,i}$ (by the Corollary to Lemma 3.2), and take $i\underline{\nu} = \Lambda(\nu^{*}_{i,i}, i\underline{u})$. The previously defined $j\underline{u}$ and $j\underline{\nu}$ are subdivided, but retain their defining properties using the same jh. It remains to define ih.

Say $\partial \nu^{*}_{i,i} < \nu^{*}_{i,i}$ is the induced subdivision of $\partial \nu_{i,i}$. We have already defined the flag subdivision

$b = \partial h \cup f'' : (iu_{,i}) \vdash (\partial \nu^{*}_{i,i} \cup i\xi''_{i,i}) \to (\xi_{,i} \times I) \vdash (\partial(\xi_{i,i} \times I) \cup \xi_{i,i} \times 0)$

(where $i\underline{\xi}'' = i\underline{u}\langle\underline{\xi}''\rangle$). We also have the restriction of f' to a flag subdivision $c : (iu_{,i}) \vdash i\xi'_{i,i} \to (\xi_{,i} \times I) \vdash \xi_{i,i} \times 1$. b and c agree over $\partial(i\xi'_{i,i})$, by construction of ∂h. Now, using the obvious isomorphism of $(\xi_{,i} \times I) \vdash \xi_{i,i} \times 0 \approx (\xi_{,i}) \vdash \xi_{i,i}$, b expresses $(\xi_{,i}) \vdash \xi_{i,i}$ as $(p\circ f'')^{*}(\eta_{,i}) \vdash \eta_{i,i}$. Similarly c expresses $(\xi_{,i}) \vdash \xi_{i,i}$ as $(p\circ f')^{*}(\eta_{,i}) \vdash \eta_{i,i}$. Since $p\circ f'' = f = p\circ f'$, these expressions are equal. It follows from the Appendix, section I, 1 that b and c extend to a flag subdivision $d : (iu_{,i}) \vdash \nu^{*}_{i,i} \to (\xi_{,i} \times I) \vdash \xi_{i,i} \times I$. $b \cup c$ is in fact given by subdivisions of $(n+1, n)$-stratifications.

$f'' \cup \partial h \cup f' : \Lambda(i\,\xi''_{i,i} \cup \partial \nu^*_{i,i} \cup i\,\xi'_{i,i'}, i\,\underline{\mu}) \to$

$$\Lambda(\xi_{i,i} \times 0 \cup \partial \xi_{i,i} \times I \cup \xi_{i,i} \times 1, \underline{\xi} \times I).$$

The Corollary to Lemma 3.2 shows that we may choose d to come from a subdivision of an (n+1, n)-stratification subdivision, which we call ih, $ih : \Lambda(\nu^*_{i,i'}, i\,\underline{\mu}) \to \Lambda(\xi_{i,i} \times I, \underline{\xi} \times I)$.

This completes the inductive step. At the end of the induction we have a subdivision $\underline{\mu}^{\vee}$ of $\underline{\eta} \times C \times I$, a restriction $\underline{\nu}^{\vee}$, defined as the subdivision induced by $\underline{\mu}^{\vee}$ of $\cup\{i\,\underline{\nu} : i = 0, \ldots, n\}$ and a subdivision $h = \cup\{ih\}$ $h : \underline{\nu} \to \underline{\xi} \times I$. They form a structure for the required concordance between g'' and g'.

Lemma 5.3. Given stratified maps $f, g, h : \underline{\xi} \mapsto \underline{\eta}$ such that f is concordant to g, and g to h. Then f is concordant to h.

Let $(D^u, B, (\underline{\eta} \times B \times I)'', (\underline{\xi} \times I)'', F'')$ be a structure for a concordance between f and g; and let $(D^t, C, (\underline{\eta} \times C \times I')', (\underline{\xi} \times I')', G')$, where $I' = [2, 3]$ be a concordance between g and h. Let A be a cell complex triangulation of a disk D^s with s large. Let a, b, c be vertices of A, B, C. The proof of Lemma 5.2 shows that the structures for g given by

$(\underline{\xi} \times 1)'' \times c \times a < (\underline{\eta} \times B \times 1)'' \times C \times A$ and

$(\underline{\xi} \times 2)' \times b \times a < (\underline{\eta} \times C \times 2)' \times B \times A$ extend to a concordance $(\underline{\xi} \times I^*)^* < \underline{\zeta}^*$ where $I^* = [1, 2]$ and $\underline{\zeta}^*$ is a subdivision of $\underline{\eta} \times B \times C \times A \times I^*$ which restricts to subdivisions of $(\underline{\eta} \times B \times 1)'' \times C \times A$ and $(\underline{\eta} \times C \times 2)' \times B \times A$. Then $(\underline{\eta} \times B \times I)'' \times C \times A \cup \underline{\zeta}^* \cup (\underline{\eta} \times C \times I')' \times B \times A$ has a restriction which subdivides $(\xi \times I)'' \cup (\underline{\xi} \times I^*)^* \cup (\underline{\xi} \times I)'$. This pair defines a concordance between f and h.

Lemma 5.4. Let $f : \underline{K} \mapsto \underline{L}$ be a map of n-stratifications and $\underline{\eta}/\underline{L}$ an (n+1, n)-stratification. Then there is an induced (n+1, n)-stratification $f^*\underline{\eta}/\underline{K}$ such that f extends to a map of stratifications $f^* : f^*\underline{\eta} \mapsto \underline{\eta}$. $f^*\underline{\eta}$ is unique up to isomorphism mod $\underline{K}$. If f is concordant to f_1, then $f^*\underline{\eta} \approx f_1^*\underline{\eta}$ mod $\underline{K}$.

Let $(D^u, B, (\underline{L} \times B)'', \underline{K}'', f'')$ be a structure for f. Then there is a subdivision $\underline{\eta}''/(\underline{L} \times B)''$ of $\underline{\eta} \times B/\underline{L} \times B$ (by Proposition 3.7), which is unique up to isomorphism mod$(\underline{L} \times B)''$. $\underline{\eta}'' \upharpoonright \underline{K}''$ can be uniquely amalgamated over $\underline{K}$, using f'', to give a stratification which we call for now $f^*(\underline{\eta}, (\underline{L} \times B)'')/\underline{K}$. Clearly f extends to a map of stratifications $f^* : f^*(\underline{\eta}, (\underline{L} \times B)'') \mapsto \underline{\eta}$. If $(D^t, C, (\underline{L} \times C)', \underline{K}', f')$ is another structure for f, then the proof of Lemma 5.2 shows that if D^s, A, a, b and c are defined as in the proof of Lemma 5.3, then there are: a subdivision $\underline{H}^*$ of $\underline{L} \times B \times C \times A \times I$ which restricts to subdivisions $\underline{H}''$ of $(\underline{L} \times B)'' \times C \times A \times 0$ and $\underline{H}'$ of $(\underline{L} \times C)' \times B \times A \times 1$; a restriction $\underline{J}^* < \underline{H}^*$ which restricts to subdivisions of $\underline{K}'' \times c \times a \times 0$ and $\underline{K}' \times b \times a \times 1$; and a subdivision $f^* : J^* \to \underline{K} \times I$ which extends $f'' \times 0$ and $f' \times 1$. Our construction yields an (n+1, n)-stratification $F^*(\underline{\eta}, \underline{H}^*)/\underline{K} \times I$, which extends $f^*(\underline{\eta}, \underline{H}'')/\underline{K}$ and $f^*(\underline{\eta}, \underline{H}')/\underline{K}$. By Proposition 3.4, $F^*(\underline{\eta}, \underline{H}^*) \approx f^*(\underline{\eta}, \underline{H}'') \times I$. Hence $f^*(\underline{\eta}, \underline{H}'') \approx f^*(\underline{\eta}, \underline{H}')$ mod $\underline{K}$. By inspection, $f^*(\underline{\eta}, \underline{H}'') = f^*(\underline{\eta}, (\underline{L} \times B)'')$, and

$f^*(\underline{\eta}, \underline{H}') = f^*(\underline{\eta}, (\underline{L} \times C)')$. So $f^*(\underline{\eta}, (\underline{L} \times B)'') \approx f^*(\underline{\eta}, (\underline{L} \times C)')$; and $f^*\underline{\eta}$ does not depend on the structure chosen for f.

That $f^*\underline{\eta}$ is independent of the choice of f within its concordance class follows from Proposition 3.4, as used above.

Lemmas 5.1 and 5.3 show we can define a category $\mathcal{A}$ whose objects are n-stratifications $\underline{K}$, and whose morphisms are concordance classes of stratified maps. Lemma 5.4 shows there is a contravariant functor $\varphi : \mathcal{A} \to$ Sets defined by: $\varphi(\underline{K})$ is the set of isomorphism classes mod $\underline{K}$ of $(n+1, n)$-stratifications over $\underline{K}$; and if $f : \underline{K} \mapsto \underline{L}$, then $\varphi(f)\underline{\eta} = f^*\underline{\eta}$. (We leave the reader to check, using Lemma 5.1, that $\varphi(g \circ f) = \varphi(f) \circ \varphi(g)$.) $\mathrm{Mor}[\underline{K}, \underline{L}]$ will denote the set of morphisms from $\underline{K}$ to $\underline{L}$ in $\mathcal{A}$.

Our goal is to show that, under some restrictions, φ is representable. The first restriction is that we shall only deal with $(n+1, n)$-stratifications $\underline{\xi}/\underline{K}$ such that $|\xi_{n+1,}|$ is a manifold and $(|K_{n,}|, \ldots, |K_{o,}|)$ the intrinsic filtration of $|K_{n,}|$ in $|\xi_{n+1,}|$. An equivalent condition is that for every cell σ of $\underline{K}$, say $\sigma \in K_{i,i}$; then $|\xi(\sigma)|$ is a disk and $(|K_{n,i}(\sigma)|, \ldots, |K_{i+1,i}(\sigma)|, \sigma)$ is the intrinsic filtration of $|K_{n,i}(\sigma)|$ in $|\xi(\sigma)|$ (compare the discussion on p. 4.1 ff.). The condition that $|\xi_{n+1,}|$ be a manifold is important. The condition on $(|K_{n,}|)$ is not essential, but has the advantage that we have only to keep track of $|\xi_{n+1,}|$ and $|K_{n,}|$; the other $|K_{i,}|$ for $i = 0, \ldots, n-1$ are automatically accounted for. This restriction will be in force during the rest of this paper (unless otherwise stated).

The second restriction is minor; it arises because our definition of polyhedra requires them to be finite-dimensional. To make a precise statement requires some more definitions:

A family of link pairs $\mathcal{L}$ is a countable set of (p.ℓ. isomorphism classes of) pairs of compact polyhedra (G, F) such that:

link pairs 1 if $(G, F) \in \mathcal{L}$, then (G, F) is not a cone- or suspension-pair;

link pairs 2 if $(G, F) \in \mathcal{L}$ and $x \in F$, then $b.\ell k(x; G, F) \in \mathcal{L}$ (see Chapter 1, §2 for the definition of "basic link"). A family of links $\mathcal{F}$ is a set $\{F\}$ of polyhedra such that $\{(F, F)\}$ is a family of link pairs.

A pair of polyhedra (Y, X) is a $\mathcal{L}$-pair if for every $x \in X$, $b.\ell k(x; Y, X) \in \mathcal{L}$. Similarly one defines when X is a $\mathcal{F}$-polyhedron. An (n+1, n)-stratification $\underline{\xi}/\underline{K}$ is a $\mathcal{L}$-stratification if for every $\sigma \in K_{i,i}$, the fibre $(G, F_n, \dots, F_{i+1})$ of $(\xi, K_{,i}) \restriction \langle\sigma\rangle$ is such that $(G, F_n) \in \mathcal{L}$. ($(F_n, \dots, F_{i+1})$ is the intrinsic filtration of F_n in G by the convention we just made.)

It follows that $(|\underline{\xi}|, |K_{n,}|)$ is a $\mathcal{L}$-pair and that $(|K_{n,}|, \dots, |K_{o,}|)$ is the intrinsic filtration of $|K_{n,}|$ in $|\underline{\xi}|$. Conversely, the proof of Theorem 2.1 shows that if (Y, X) is a $\mathcal{L}$-pair, then X has a $\mathcal{L}$-stratification in Y with respect to the intrinsic filtration of X in Y.

If (Y, X) is a pair of polyhedra, then we define $\mathcal{L}(Y, X)$ as the smallest family of link pairs such that (Y, X) is a $\mathcal{L}(Y, X)$-pair. $\mathcal{L}(Y, X)$ is constructed by induction on dim Y: up to p.ℓ. isomorphism, there are at most countably many pairs of the form $b.\ell k(x; Y, X)$ as x varies in X.

Each $\mathcal{L}(b.\ell k(x; Y, X))$ exists by inductive hypothesis and is in fact finite, since $b.\ell k(x; Y, X)$ is compact. So $\mathcal{L}(Y, X) = \{b.\ell k(x; Y, X)\} \cup \cup \{\mathcal{L}(b.\ell k(x; Y, X))\}$ is indeed a family of link pairs.

A similar argument shows that if $\mathcal{L}$ is a family of link pairs, then there is a well-defined family of links $\mathfrak{F}(\mathcal{L})$ defined to be the smallest family such that for every $\mathcal{L}$-pair (Y, X), X is an $\mathfrak{F}(\mathcal{L})$-polyhedron.

The fact that $\mathcal{L}(G, F)$ is finite if G, F are compact enables one to order the elements of any family of link pairs $\mathcal{L}$, say $(G^1, F^1), (G^2, F^2), \ldots$ so that for each k, $\{(G^j, F^j) : j = 1, \ldots, k\} = \mathcal{L}(k)$ is a family of link pairs. Such an ordering will be called a <u>standard ordering</u> of $\mathcal{L}$.

$\mathcal{L}$ is <u>finite-dimensional</u> if there is an integer k such that for every $(G, F) \in \mathcal{L}$, $\dim G \leq k$. $\mathcal{L}$ is of <u>codimension $\geq q$</u> (or of <u>codimension q</u>) if for every $(G, F) \in \mathcal{L}$, $\dim F \leq \dim G - q$ (or $\dim F = \dim G - q$, respectively). A family of links $\mathfrak{F}$ is of <u>constant local dimension</u> if for every $F \in \mathfrak{F}$, $\dim st(x, F)$ is constant as x varies in F.

A pair (Y, X) is <u>boundary-less</u> if $\text{bdy } Y = \emptyset = \text{bdy } X$ and neither Y nor X is a single point. A family of link pairs $\mathcal{L}$ is <u>boundary-less</u> if every (G, F) in $\mathcal{L}$ is boundary-less. Observe that these three conditions on a pair (Y, X) are equivalent:

1. $\mathcal{L}(Y, X)$ is boundary-less;
2. for every $x \in X$, $b.\ell k(x; Y, X)$ is boundary-less;
3. $\text{bdy}_Y X = \text{bdy } X$, and if Y' is a regular neighbourhood of X in Y, then $(\text{bdy } Y', \text{bdy } X) < \text{bdy}(Y', X)$.

Now let $\underline{K}$ be an $\mathfrak{F}$-n-stratification, where $\mathfrak{F}$ is boundary-less. Then $|K_{n,o}|$ is a weak regular neighbourhood of $|K_{o,o}|$. From the foregoing we see that $|K_{o,o}| \cap \text{bdy}\, |K_{n,o}| = \text{bdy}\, |K_{o,o}|$ which is covered by a subcomplex bdy $K_{o,o} < K_{o,o}$ (by Theorem 1.6). And bdy $|K_{n,o}| = |\Lambda(\text{bdy}\, K_{o,o}, \underline{K})| \cup |\mathcal{K}_o^*(\underline{K})|$. Hence, by induction on n, one proves that bdy $|K_{n,}|$ is covered by a restriction bdy $\underline{K} < \underline{K}$; and then we have that $|(\text{bdy}\, K)_{n,}| < \text{bdy}\, |K_{n,}|$.

The most important example of a family of link pairs is this: Let $\mathfrak{F}$ be a finite-dimensional, boundary-less family of links of constant local dimension, and let $q \geq 3$. Then $\mathfrak{s}^q(\mathfrak{F})$ is the family of link pairs consisting of all pairs (S, F), where S is a sphere, $F \in \mathfrak{F}$, and dim F = dim S - q. Observe that (Y, X) is an $\mathfrak{s}^q(\mathfrak{F})$-pair for some such family $\mathfrak{F}$ if and only if:

X is of constant local dimension;

bdy X is collared in X;

in a neighbourhood of X, Y is a manifold of constant local dimension q + dim X;

bdy Y $\cap$ X = bdy X, and (bdy Y, bdy X) < bdy(Y, X).

Since $q \geq 3$, we also have that the intrinsic variety of X in Y is just the intrinsic variety of X.

Let $\mathfrak{J}$ be a family of link pairs, r a positive integer. Then a $\mathfrak{J}$-stratification $\underline{r\gamma}/\underline{rU}$ is <u>r-universal</u> if for every $\mathfrak{F}(\mathfrak{J})$-stratification $\underline{K}$ with dim $|\underline{K}| \leq r$, the function φ: Mor[$\underline{K}$, $\underline{rU}$] → the set of isomorphism classes modulo $\underline{K}$ of $\mathfrak{J}$-stratifications over $\underline{K}$, defined by $\varphi(f) = f^*(\underline{r\gamma})$, is a bijection.

Observe that if $\underline{K}$ is an n-stratification and $\underline{K}'$ a subdivision of $\underline{K}$, then the identity map $id : \underline{K}' \mapsto \underline{K}$ is an isomorphism in $\mathcal{A}$. Also if B is a cell complex triangulation of a disk, then the projection $p : \underline{K} \times B \mapsto \underline{K}$ is an equivalence in $\mathcal{A}$ in the sense that p induces bijections $Mor[\underline{J}, \underline{K}{\times}B] \longleftrightarrow Mor[\underline{J}, \underline{K}]$ and $Mor[\underline{K}, \underline{L}] \longleftrightarrow Mor[\underline{K}{\times}B, \underline{L}]$ for any n-stratifications $\underline{J}$ and $\underline{L}$. Hence if $\underline{r\gamma}/\underline{rU}$ is an r-universal $\mathcal{L}$-stratification, then any subdivision $\underline{r\gamma}'/\underline{rU}'$ of $\underline{r\gamma}{\times}B/\underline{rU}{\times}B$ is also r-universal.

Proposition 5.5. Let $\mathcal{L}$ be a finite, boundary-less family of link pairs of codimension ≥ 3 such that for every $(G, F) \in \mathcal{L}$, G is a sphere. Then for each r, there is an r-universal $\mathcal{L}$-stratification.

Take a standard ordering of $\mathcal{L}$, and set $\mathcal{L}(k) = \{(G^j, F^j) : j = 1, \ldots, k\}$. We use induction on k: Assume that for each integer s there is an s-universal $\mathcal{L}(k)$-stratification $\underline{\gamma}(k)/\underline{U}(k)$ such that $|\underline{\gamma}(k)|$ (which is a manifold) is connected and has trivial tangent cS-block bundle $\tau\ |\underline{\gamma}(k)|$. Our task is to construct, for each r, an r-universal $\mathcal{L}(k+1)$-stratification satisfying the analogous conditions. We write (G, F) for (G^{k+1}, F^{k+1}), and let $(F_n, \ldots, F_1)$ be the intrinsic filtration of F in G.

Note that (G^1, F^1) must be $(S^m, \emptyset)$ for some $m \geq 2$ (since $\dim \emptyset = -1$). Hence a $\mathcal{L}(1)$-stratification is just a cS^m-block bundle over a manifold. Now $r\gamma/rU$ exists by Rourke and Sanderson's [19, I,

§ 2], and $|r\gamma|$ is connected. The argument below gives some details; including how to arrange that $\tau|r\gamma|$ is trivial. We proceed to the general case, $\mathcal{J}(k+1)$:

First we construct a $c(G, F_n, \ldots, F_1)$-flag $(\xi', \mu'_n, \ldots, \mu'_1)/H'$, where H' is a simplicial complex, which is r-universal in this sense: for every simplicial complex P of dimension $\leq r$, the function $\psi : [|P|, |H|] \rightarrow$ the set of isomorphism classes of $c(G, F_n, \ldots, F_1)$-block bundles over P, defined by $\psi(h) = h^*(\xi', \mu')$, is a bijection. This is done using Rourke and Sanderson's [22, II]. Let $P\ell^{\sim}(G, F)$ be the Δ-group which has for a-simplexes all p.ℓ. automorphisms $f : \Delta^a \times cG \rightarrow \Delta^a \times cG$ (Δ^a is the a-simplex) such that:

f is the identity on $\Delta^a \times c$;

$f^{-1}(\Delta^a \times cF) = \Delta^a \times cF$

$f^{-1}(\Delta^a \times G) = \Delta^a \times G$;

$f^{-1}(\Delta^b \times cG) = \Delta^b \times cG$ for every face $\Delta^b < \Delta^a$. Since F is of codimension at least 3 in G, the intrinsic filtration of F in G is $(F_n, \ldots, F_1)$. Hence by Lemma 1.10, $f^{-1}(\Delta^a \times F_i) = \Delta^a \times F_i$, for $i = 1, \ldots, n-1$. It follows that a $P\ell^{\sim}(G, F)$-block bundle θ over a simplicial complex P, in the sense of Rourke and Sanderson, can be given canonically the structure of a $c(G, F_n, \ldots, F_1)$-flag $(\theta, \nu_n, \ldots, \nu_1)/P$; and conversely, given such a flag $(\theta, \nu)/P$ with P a simplicial complex, then θ/P has canonically a $P\ell^{\sim}(G, F)$-block bundle structure. The existence of an r-universal $c(G, F_n, \ldots, F_1)$-flag now follows from [22, II, Corollary 2.6] applied to $P\ell^{\sim}(G, F)$.

We can replace $(\xi',\mu')/H'$ by another r-universal flag $(\xi'',\mu'')/H''$ such that $|H''|$ is a connected manifold without boundary. Then $|\xi''|$ is a connected manifold. Let $\nu/|\xi''|$ be its stable normal cS-block bundle. Let H be a simplicial triangulation of $|\nu \restriction |H''|| - \nu^{\bullet} \restriction |H''|$ (regarded as a polyhedron, and not as a subpolyspace of $|\nu|$). Then we can choose an r-universal flag $(\xi,\mu)/H$, and now $|H|$ is a manifold without boundary, and $|\xi|$ is a connected manifold with boundary $\xi^{\bullet}$, whose tangent bundle, $\tau|\xi|$, is trivial.

Let $\underline{\tau}/\underline{L}$ be an $(n+1,n)$-bs completion of $(\mu_n,\ldots,\mu_1)/H$ in (ξ,μ) (see Lemma 2.2). Thus $\Lambda_o^*(\underline{\eta})$ consists of the blocks added to (ξ,μ) to form $\underline{\tau}$. Since (G, F) is not a cone- or suspension-pair, $\Lambda_o^*(\underline{\eta})$ is a $\mathcal{S}(k)$-stratification. Pick an integer $s \geq r+1$ and $\geq \dim |\Lambda_o^*(\underline{L})| = \dim \mu_n^{\bullet}$, and let $\underline{\gamma}(k)/\underline{U}(k)$ be an s-universal $\mathcal{S}(k)$-stratification such that $\tau|\underline{\gamma}(k)|$ is trivial. Let $h : \Lambda_o^*(\underline{L}) \mapsto \underline{U}(k)$ classify $\Lambda_o^*(\underline{\eta})$. Pick a structure $(D^u, E, (\underline{U}(k)\times E)', \Lambda_o^*(\underline{L})', h')$ for h such that $\dim|\underline{\gamma}(k)| + u \geq 2 \dim \xi^{\bullet} + 2$. Set $\underline{U}^* = (\underline{U}(k)\times E)'$, and let $\underline{\gamma}^*/\underline{U}^*$ be a subdivision of $\underline{\gamma}(k) \times E$; then $\underline{\gamma}^*$ is again an s-universal $\mathcal{S}(k)$-stratification with trivial tangent bundle. Note that h' extends to a subdivision $h^* : \underline{\gamma}^* \restriction \Lambda_o(\underline{L})' \rightarrow \underline{\eta}$. We write $\Lambda_o^*(\underline{\eta})'$ for $\underline{\gamma}^* \restriction \Lambda_o^*(\underline{L})'$.

Since $\mathcal{S}(k)$ is boundary-less, $\text{bdy}|\underline{U}^*|$ is covered by a restriction $\text{bdy}\,\underline{U}^* < \underline{U}^*$. Let $\underline{U}^{\sim} = \underline{U}^* \cup (\text{bdy}\,\underline{U}^* \times I)$ with $\text{bdy}\,\underline{U}^*$ identified to $\text{bdy}\,\underline{U}^* \times 0$; similarly let $\underline{\gamma}^{\sim} = \underline{\gamma}^* \cup (\underline{\gamma} \restriction \text{bdy}\,\underline{U}^*) \times I$ with $\underline{\gamma}^* \restriction \text{bdy}\,\underline{U}^*$ identified to $(\underline{\gamma}^* \restriction \text{bdy}\,\underline{U}^*) \times 0$. Then $\underline{\gamma}^{\sim}/\underline{U}^{\sim}$ is an s-universal $\mathcal{J}(k)$-stratification, since the inclusion $i : \underline{Y}^* \mapsto \underline{Y}^\sim$ and the retraction $r : \underline{Y}^\sim \mapsto \underline{Y}^*$ which is the identity on $\underline{Y}^*$ and projects $(\underline{Y}^* \setminus \text{bdy}\, \underline{U}^*) \times I$ to $(\underline{Y}^* \setminus \text{bdy}\, \underline{U}^*) \times 0$ induce inverse functions $\text{Mor}[\underline{J}, \underline{U}^*] \longleftrightarrow \text{Mor}[\underline{J}, \underline{U}^\sim]$ for all $\underline{J} \in \mathcal{A}$. The advantage of using $\underline{U}^\sim$ is that now $\Lambda_o^*(\underline{L})' \cap \text{bdy}\, \underline{U}^\sim = \emptyset$. It follows that the manifold $|\Lambda_o^*(\underline{\eta})'|$ is a proper submanifold of $|\underline{Y}^\sim|$; that is, $\text{bdy}|\underline{Y}^\sim| \cap |\Lambda_o^*(\underline{\eta})'| = \text{bdy}|\Lambda_o^*(\underline{\eta})'|$.

Now both $|\underline{Y}^\sim|$ and $|\Lambda_o^*(\underline{\eta})'|$ (which is p.ℓ. isomorphic to $|\Lambda_o^*(\underline{\eta})|$, which is itself $< \text{bdy}|\underline{\xi}|$) have trivial tangent bundles, and we have arranged that $\dim|\underline{Y}^\sim| \geq 2 \dim|\Lambda_o^*(\underline{\eta})'| + 1$. Hence $|\Lambda_o^*(\underline{\eta})'|$ has a cS-block bundle neighbourhood θ/P in $|\underline{Y}^\sim|$ which is stable, and whose fibre has the same dimension, z say, over every cell of P, since $\underline{\xi}^\cdot$ is connected. By Rourke and Sanderson's [19, II, Corollary 5.6], $(\tau|\underline{Y}^\sim|) \setminus |\Lambda_o^*(\underline{\eta})'| = \tau|\Lambda_o^*(\underline{\eta})'| \oplus \theta$. It follows from their [19, II, Corollary 5.3] that trivializations of $\tau|\underline{Y}^\sim|$ and $\tau|\underline{\xi}|$ induce a trivialization of θ, $\alpha : \theta \approx \epsilon^z/P$. By subdividing $\underline{Y}^\sim$ if necessary we may find a restriction $\underline{\lambda}/\underline{W}$ of $\underline{Y}^\sim/\underline{U}^\sim$ such that $|\underline{\lambda}|$ is a regular neighbourhood of $|\Lambda_o^*(\underline{\eta})'|$ in $|\underline{Y}^\sim|$. Then α induces a p.ℓ. isomorphism $\alpha_o : |\Lambda_o^*(\underline{\eta})'| \times (D^z, c) \to (|\underline{\lambda}|, |\Lambda_o^*(\underline{\eta})'|)$, where $c \in \text{int}\, D^z$ and $|\Lambda_o^*(\underline{\eta})'| \times c \to |\Lambda_o^*(\underline{\eta})'|$ is the obvious map. The following lemma shows that there is an isotopy of α_o rel $|\Lambda_o^*(\underline{\eta})'| \times c$ to some α_1 which corresponds $|\Lambda_o^*(\underline{\eta})'_{i,}| \times D^z \longleftrightarrow |\lambda_{i,}|$ for all i.

Lemma 5.6(n). Given (n+1, n)-stratifications $\underline{\eta} < \underline{\xi}$ such that $|\underline{\xi}|$, $|\underline{\eta}|$ are manifolds, $|\eta_{i,i}|$ is a proper submanifold of $|\xi_{i,i}|$ for

$i = 0, \ldots, n$, and $\dim|\xi_{i,i}| \geq 6$ for all i. Given a p.ℓ. isomorphism $a : |\xi_{n+1,}| \to |\eta_{n+1,}| \times D^k$ with $|\eta_{n+1,}|$ corresponding to $|\eta_{n+1,}| \times c$ in the obvious way (where D^k is a k-disk, $c \in \operatorname{int} D^k$). Then there is an isotopy a_t of a rel$|\eta_{n+1,}|$ such that a_1 is a p.ℓ. isomorphism $a_1 : (|\xi_{n+1,}|, \ldots, |\xi_{o,}|) \to (|\eta_{n+1,}|, \ldots, |\eta_o|) \times D^k$.

We use induction on n. Lemma 5.6(-1) is vacuous. Now assume Lemma 5.6(n-1): By uniqueness of regular neighbourhoods we can isotop a rel$|\eta_{n+1,}|$ to some a' which restricts to a p.ℓ. isomorphism of pairs $a' : (|\xi_{n+1,0}|, \dot{\xi}_{n+1,0}) \to (|\eta_{n+1,0}|, \dot{\eta}_{n+1,0}) \times D^k$. The diagram

$$\begin{array}{ccccc}
|\xi_{0,0}| & \hookrightarrow & |\xi_{n+1,0}| & \xrightarrow{\;a'\;} & |\eta_{n+1,0}| \times D^k \\
\Big\uparrow \text{inclusion} & & \Big\uparrow \text{inclusion} & & \Big\uparrow \text{id} \times \text{inclusion} \\
|\eta_{0,0}| & \hookrightarrow & |\eta_{n+1,0}| & \xrightarrow[\text{obvious}]{} & |\eta_{n+1,0}| \times c
\end{array}$$

commutes, and the horizontal maps and right-hand vertical map are simple homotopy equivalences. It follows that $|\eta_{o,o}| \hookrightarrow |\xi_{o,o}|$ is a simple homotopy equivalence of manifolds. Since $|\eta_{o,o}|$ is a proper submanifold of $|\xi_{o,o}|$, $|\eta_{o,o}|$ has a normal cS^{k-1}-block bundle neighbourhood in $|\xi_{o,o}|$. It follows from the s-cobordism theorem (since $\dim|\xi_{o,o}| \geq 6$) that $|\xi_{o,o}|$ has the structure of a cS^{k-1}-block bundle ν/P over $|\eta_{o,o}|$. Now $(\xi_{,o})/\xi_{o,o}$ is a blocking of the block decomposition $[\nu; (\eta_{,o})]/\eta_{o,o}$. Let $\nu^\wedge/P^\wedge$ be a complementary blocking. Then $|\xi_{n+1,0}|$ has two

cS^{k-1}-block bundle structures over $|\eta_{n+1,0}| : \nu^{\wedge}/P^{\wedge}$ and $a'^{-1}(\epsilon^k/|\eta_{n+1,0}|)$, where ϵ^k denotes the trivial block bundle. Since $|\eta_{n+1,0}|$ is a manifold, these are isotopic (by Rourke and Sanderson's [19, I, Theorem 4.4]). Hence $\nu^{\wedge}$ is trivial, and so ν is trivial. Thus a' is isotopic rel $|\eta_{n+1,}|$ to an isomorphism a'' which restricts to $a'' : (|\xi_{n+1,0}|, |\xi_{n,0}|, \ldots, |\xi_{0,0}|) \to (|\eta_{n+1,0}|, \ldots, |\eta_{0,0}|) \times D^k$ and $a'' : \dot{\xi}_{n+1,0} \to \dot{\eta}_{n+1,0} \times D^k$.

Then a'' restricts to p.ℓ. isomorphisms $b^{\bullet} : |\Lambda_o^*(\underline{\xi})| \to |\Lambda_o^*(\underline{\eta})| \times D^k$ and $b : |\mathcal{C}_o(\underline{\xi})| \to |\mathcal{C}_o(\underline{\eta})| \times D^k$. By Lemma 5.6(n-1), there is an isotopy b_t of b rel $|\mathcal{C}_o(\underline{r})|$ to a p.ℓ. isomorphism $b_1 : (|\mathcal{C}_o(\underline{\xi})_{n+1,}|, \ldots, |\mathcal{C}_o(\underline{\xi})_{1,}|) \to (|\mathcal{C}_o(\underline{\eta})_{n+1,}|, \ldots, |\mathcal{C}_o(\underline{\eta})_{1,}|) \times D^k$. Since $(|\Lambda_o^*(\underline{\xi})_{n+1,}|, \ldots, |\Lambda_o^*(\underline{\xi})_{1,}|, |\Lambda_o^*(\underline{\eta})_{n+1,}|) <$ $<$ bdy $(|\mathcal{C}_o(\underline{\xi})_{n+1,}|, \ldots, |\mathcal{C}_o(\underline{\xi})_{1,}|, |\mathcal{C}_o(\underline{\eta})_{n+1,}|)$, we may choose b_1 to extend $b^{\bullet}$. Then a_1, defined to be a'' on $|\xi_{n+1,0}|$ and b_1 on $|\mathcal{C}_o(\underline{\xi})_{n+1,}|$, is the required p.$\ell$. isomorphism. This completes the inductive step, and Lemma 5.6 is proved.

To return to the proof of Proposition 5.6: By Theorem 3.1, we can further isotop $\alpha_1 : |\Lambda_o(\underline{\eta})'| \times D^z \to |\underline{\lambda}|$ through isomorphisms of (n+1, n)-ns's to an isomorphism of stratifications $\alpha^{\wedge} : \Lambda_o^*(\underline{\eta})^{\wedge} \to \underline{\lambda}^{\wedge}$, where $\Lambda_o^*(\underline{\eta})^{\wedge}$ is a subdivision of $\Lambda_o^*(\underline{\eta})' \times A$, A is a cell complex triangulation of D^z, and $\underline{\lambda}^{\wedge}$ is a subdivision of $\underline{\lambda}$. Let $\underline{\eta}^{\wedge} = \underline{\eta} \times A + \Lambda_o^*(\underline{\eta})^{\wedge}$; then $\underline{\eta}^{\wedge}$ is a completion of $(\xi, \mu)\times A/H\times A$. This is still an r-universal c(G, F)-flag, which we re-name $(\xi, \mu)/H$. Finally define

$r\underline{\gamma} = (\underline{\gamma}^{\sim} + \underline{\lambda}^{\wedge}) \times I \cup \underline{\eta}^{\wedge}$, with $\Lambda_o^*(\underline{\eta})^{\wedge}$ identified to $\underline{\lambda}^{\wedge} \times 0$ by $\alpha^{\wedge}$. Then $r\underline{\gamma}$ is a $\mathcal{B}(k+1)$-stratification. Also $\tau|(\underline{\gamma}^{\sim}+\underline{\lambda}^{\wedge}) \times I|$ and $\tau|\underline{\eta}^{\wedge}|$ are both trivial; and the choice of $\alpha^{\wedge}$ ensures that they have trivializations which agree over $|\underline{\lambda}^{\wedge}|$. Hence $\tau|r\underline{\gamma}|$ is trivial. It remains to show that $r\underline{\gamma}$ is r-universal:

Proof that φ is onto: Let $\underline{\xi}/\underline{K}$ be a $\mathcal{B}(k+1)$-stratification with $\dim|\underline{K}| \leq r$. Let $P = \{\sigma \in \text{some } K_{i,i} : (\xi(\sigma), K_{n,i}(\sigma))$ has fibre $(G, F)\}$. P need not be $K_{o,o}$; but by taking a filtration of $(|K_{n,}|, \ldots, |K_{o,}|)$ equivalent to the given one, and using the corresponding equivalent stratification to $\underline{\xi}$, we can arrange that $P = K_{o,o}$. Similarly we shall assume $rU_{o,o} = H$. By subdividing $\underline{\xi}$ if necessary (by the Corollary to Lemma 3.2) we may assume P is a simplicial complex. Since $\dim P \leq r$ there is a classifying map $p : |P| \to |H|$ for the flag $(\xi, K_{,o})/P$. Let $((H\times B)'', P'', p'')$ be a structure for p regarded as a 0-bs map. There are subdivisions $\underline{\xi}''/\underline{K}''$ of $\underline{\xi}$ such that $K''_{o,o} = p''(P'')$, and $r\underline{\gamma}''/r\underline{U}''$ of $r\underline{\gamma} \times B/r\underline{U} \times B$ such that $rU''_{o,o} = (H\times B)''$. Then, by definition of induced flags, p''^{-1} extends to an isomorphism of flags, $p^{\wedge} : (\xi'', K''_{,o}) \approx (\mathcal{J}'', \mu'')|P''$. The method of proof of Theorem 3.1 shows that we can choose $p^{\wedge}$ (after subdividing $\Lambda_o^*(\underline{\xi}'')$ and $\Lambda_o^*(r\underline{\gamma}'')$ if necessary) to be an isomorphism of stratifications $p^{\wedge} : \Lambda(P'', \underline{\xi}) \approx \Lambda(\underline{P}'', r\underline{\gamma}'')$ (this uses the fact that $p^{\wedge}$ is now an inclusion of one flag in another). Set $\underline{\gamma}'' = \mathcal{C}_o(r\underline{\gamma}'')$; then $\underline{\gamma}''$ is an s-universal $\mathcal{B}(k)$-stratification. By definition, the restriction of $p^{\wedge}$ to $p^* : \Lambda_o^*(\underline{K}'') \mapsto \underline{U}''$ is a classifying map for $\Lambda_0^*(\underline{\xi}'')$.

By subdividing $\mathcal{C}_0(\underline{\xi}'')$ finely enough, we may assume there is a restriction $\underline{J}'' < \mathcal{C}_0(\underline{K}'')$ such that $(|\underline{\xi}'' \vdash \underline{J}''|, |J''_{n,}|, \ldots, |J''_{1,}|)$ is a regular neighbourhood of $|\Lambda_0^*(\underline{\xi}'')|$ in $(|\mathcal{C}_0(\underline{\xi}'')|, |\mathcal{C}_0(\underline{K}'')_{n,}|, \ldots, |\mathcal{C}_0(\underline{K}'')_{1,}|)$. Since $\Lambda_0^*(\underline{\xi}'') < \text{bdy}\, \mathcal{C}_0(\underline{\xi}'')$, we may also arrange that $\underline{\xi}'' \vdash \underline{J}''$ is a subdivision of $\Lambda_0^*(\underline{\xi}'') \times I$; here we say $\underline{\xi}'' \vdash \Lambda_0^*(\underline{K}'')$ corresponds to $\Lambda_0^*(\underline{\xi}'') \times 0$, and $\underline{\xi}'' \vdash 1\underline{J}''$ to $\Lambda_0^*(\underline{\xi}'') \times 1$. Define the $(n-1)$-stratification $\mathcal{C}(\underline{J}'')$ to be $\{\beta \in \mathcal{C}_0(\underline{K}'') : |\beta| \subseteq \text{some } |\alpha|, \text{ with } \alpha \notin \underline{J}''\}$. Alternatively, $\mathcal{C}(\underline{J}'')$ is defined by the properties: $\mathcal{C}(\underline{J}'') < \mathcal{C}_0(\underline{K}'')$; $\mathcal{C}_0(\underline{K}'') = \underline{J}'' \cup \mathcal{C}(\underline{J}'')$; $1\underline{J}'' = \underline{J}'' \cap \mathcal{C}(\underline{J}'')$. Then $\underline{\xi}'' \vdash \mathcal{C}(\underline{J}'')$ is a $\mathcal{I}(k)$-stratification; let $q^\wedge : \mathcal{C}(\underline{J}'') \mapsto \underline{U}''$ classify it. We have the restriction $q^* : 1\underline{J}'' \mapsto \underline{U}''$ classifying $\underline{\xi}'' \vdash 1\underline{J}''$. By amalgamation, we may regard q^* as a stratified map $q^* : \Lambda_0^*(\underline{K}'') \mapsto \underline{U}''$. Since $\underline{\xi}''$ can be amalgamated over $\Lambda_0^*(\underline{K}'') \times I$, and p^*, q^* classify its restriction to the two ends, we have that p^* and q^* are concordant. Let $t : \Lambda_0^*(\underline{K}'') \times I \mapsto \underline{U}''$ be such a concordance. We have the subdivision $u : \underline{J}'' \mapsto \Lambda_0(\underline{K}'') \times I$. Note that $t \circ u$ extends $p^* \cup q^* : \Lambda_0^*(\underline{K}'') \cup 1\underline{J}'' \mapsto \underline{U}''$. Let the subdivision $''t : \underline{K} \to \Lambda_0^*(\underline{K}'') \times I$ be part of a structure for t. By Theorem 3.1 there is an isotopy ω_t of $|\Lambda_0^*(\underline{K}'') \times I|$ rel $|\Lambda_0^*(\underline{K}'') \times \{0, 1\}|$ such that $''t(''\underline{K})$ and $\omega_1 \circ u(\underline{J}'')$ have a common subdivision, $*\underline{K}$. Then the subdivision $\omega_1^{-1} \circ ''t : (''t)^{-1}(*\underline{K}) \to u(\underline{J}'')$ can be used in a structure for an improved t such that $t \circ u$ is a stratified map and still extends $p^* \cup q^*$. Then $v : \underline{K}'' \mapsto r\underline{U}''$, defined to be $p^\wedge$ on $\Lambda(K''_{0,0}, \underline{K}'')$, t on $\underline{J}''$, and $q^\wedge$ on $\mathcal{C}(\underline{J}'')$ is such that $v^*(r\underline{\gamma}'') = \underline{\xi}''$. Note that $v : \underline{K} \mapsto r\underline{U}''$ is by definition a stratified map. By Lemma 5.1, v is concordant to a stratified map

$v_1 : \underline{K} \mapsto r\underline{U}''$ such that, if $p : |r\underline{U}''| = |r\underline{U}| \times |B| \to |r\underline{U}|$ is the projection, then $p \cdot v_1 : \underline{K} \mapsto r\underline{U}$ is a stratified map. Clearly $(p \cdot v_1)^*(r\underline{\gamma}) = \underline{\xi}$. Thus φ is onto.

Proof that φ is 1-1: Given stratified maps $f_0, f_1 : \underline{K} \mapsto r\underline{U}$ such that $\dim|\underline{K}| \leq r$ and $f_0^*(r\underline{\gamma}) = f_1^*(r\underline{\gamma})$. Then, using a similar argument to the one just given, one finds a stratified map $F : \underline{K} \times I \mapsto r\underline{U}$ such that on $\underline{K} \times 0$, F is concordant to $f_0 \times 0$, and on $\underline{K} \times 1$, F is concordant to $f_1 \times 1$. This suffices (in view of Lemma 5.4) to prove that φ is 1-1.

This completes the inductive step. Hence Proposition 5.5 is proved.

Corollary. Proposition 5.5 still holds if $\mathcal{J}$ is finite-dimensional (and not just finite).

This time let $\mathcal{J}(k) = \{(G, F) \in \mathcal{J} : \dim G \leq k\}$; then each $\mathcal{J}(k)$ is a family of link pairs. By induction on k, assume the corollary for $\mathcal{J}(k)$. Say $\mathcal{J}(k+1) - \mathcal{J}(k) = \{(G^1, F^1), \ldots, (G^j, F^j), \ldots\}$. For each (G^j, F^j) we construct an r-universal flag $(\underline{\xi}^j, u^j)/H^j$ with fibre (G^j, F^j), as in the proof of Proposition 5.5. Observe that we can make the H^j all have the same dimension; then the $|\underline{\xi}^j|$ will all have the same dimension (since $\dim G^j = k+1$). Let $\underline{\eta}^j/\underline{L}^j$ be a completion of $(\underline{\xi}^j, u^j)$ for each j. Then the disjoint union $\underline{\eta}/\underline{L} = \cup\{\underline{\eta}^j/\underline{L}^j\}$ is a stratification; the point being that $|\underline{\eta}|$ is finite-dimensional and hence a polyhedron. Now continue with the proof of Proposition 5.5, noting that the integer z will do for all the $\underline{\eta}^j$ at once.

Remarks 1. Our requirement on n-stratifications $\underline{K}$ that $(|K_{n,}|, \ldots, |K_{0,}|)$ be the intrinsic filtration of $|K_{n,}|$ can be set aside by using more general definitions. One would define a family $\mathcal{J}$ var, each of whose elements is a pair (S, F) together with a variety filtration $(F_q, \ldots, F_0)$ of F in S. I leave the reader to state the axioms that such a $\mathcal{J}$ var should satisfy. One would also define the associated family $\mathfrak{Z}$ var ($\mathcal{J}$var), and $\mathcal{J}$ var-(n+1, n)-stratifications and $\mathfrak{Z}$ var-n-stratifications. There would then be a classifying theorem for $\mathcal{J}$ var-stratifications over a given $\mathfrak{Z}$ var-stratification. The only difference in the proof is that (ξ, μ) would not be constructed using $P\ell^{\sim}(S, F)$ but its Δ-subgroup $P\tilde{\ell}(S, F_q, \ldots, F_0)$, which consists of those $f : \Delta^a \times S \to \Delta^a \times S$ in $P\tilde{\ell}(S, F)$ such that $f^{-1}(\Delta^a \times F_i) = \Delta^a \times F_i$, for $i = 0, \ldots, q$.

2. A more promising generalization is simply to restrict the Δ-group $P\ell^{\sim}(S, F)$, by choosing (ξ, μ) to be universal for some Δ-subgroup of $P\ell^{\sim}(S, F)$. For example, one could take only those f in $P\ell^{\sim}(S, F)$ which preserve orientation of $\Delta^a \times S$. Or one could use only those f which are the identity on $\Delta^a \times F$; thus one would classify $\mathcal{J}$-(n+1, n)-stratifications over a very special kind of $\mathfrak{Z}$-n-stratification.

3. The restriction in the hypotheses of our theorem that for all $(G, F) \in \mathcal{J}$, G be a sphere, is necessary for the proof. Consider the point where we have to "fatten up" the flag (ξ, μ); this is necessary so that $(\xi, \mu) \cup \underline{Y}(k)$ does indeed define a variety. If $\xi^{\bullet}$ is not a

manifold, then I do not know that the cS-block bundle θ exists; this is related to (what I call) the main unsolved problem of stratified polyhedra (see Chapter 7). The same problem arises in the proof of Lemma 5.6: if $|\eta_{n+1,0}|$ is not a manifold, I cannot prove $\overset{\wedge}{\nu}$ and $a'^{-1}(\epsilon)$ isotopic. Even if these steps could be managed, (as they can - see Chapter 8) if θ existed and $\theta \mid u^{\bullet}$ lay in $\underline{U}(k)$ - how is one to extend θ over all of $|\xi|$ unless one knows that θ is trivial? And I do not know how to ensure this except by requiring $|\xi|$ and $|\underline{Y}(k)|$ to be manifolds so that one can use their tangent bundles.

Let X, Y be polyhedra, $(X_n, \ldots, X_o)$ and $(Y_n, \ldots, Y_o)$ variety filtrations of X and Y. A p.ℓ. map $f : X \to Y$ is <u>stratifiable</u> with respect to these filtrations if there are n-stratifications $\underline{K}$, $\underline{L}$ of (X) and (Y) such that $f : \underline{K} \mapsto \underline{L}$ is a stratified map. We write: $f : (X_n, \ldots, X_o) \mapsto (Y_n, \ldots, Y_o)$, or if they are intrinsic filtrations: $f : X \mapsto Y$.

Stratifiable maps $f_0, f_1 : (X) \mapsto (Y)$ are <u>concordant</u> if there is a stratifiable map $F : (X) \times I \mapsto (Y \times I)$ such that $F^{-1}(Y \times \{0,1\}) = X \times \{0,1\}$, and F equals $f_o \times 0$ on $X \times 0$ and $f_1 \times 1$ on $X \times 1$.

One can prove the analogues of Lemmas 5.1, 5.2 and 5.3, using Theorem 3.1 instead of Lemma 3.8. One can also prove this analogue of Lemma 5.4: Given polyhedra $Y \subset Z$ and X, a variety filtration $(X_n, \ldots, X_o)$ of X and a variety filtration $(Y_n, \ldots, Y_o)$ of Y in Z. Then any stratifiable map $f : (X) \mapsto (Y)$ induces a pair $f^*Z \supset X$ such that (X) is a variety filtration of X in f^*Z. The equivalence class (in the sense of Chapter 3) of (f^*Z, X) depends only on the equivalence class of (Z, Y) and the concordance class of f. Moreover if (Y) is the intrinsic filtration of Y in Z, then (X) is the intrinsic filtration of X in f^*Z. Hence there is a category $\mathcal{B}$ whose objects are polyhedra each with a variety filtration and whose morphisms are concordance classes of stratifiable maps.

Putting together Theorem 3.9 and the Corollary to Proposition 5.5 gives:

Theorem 5.7. Let $\mathcal{J}$ be a finite-dimensional, boundary-less family of link pairs of codimension at least 3, such that for every $(G, F) \in \mathcal{J}$, G is a sphere. Then for each r there is a $\mathcal{J}$- pair (rE, rV) which is r-universal in the sense that

φ : Mor[X, rV] $\longrightarrow$ the set of equivalence classes of $\mathcal{J}$-pairs (M, X) is a bijection whenever $\dim X \leq r$.

Here Mor[X, rV] is taken using the intrinsic filtrations of X and rV. We also note that rE and every M are manifolds.

The most important case of this theorem is when $\mathcal{J}$ is of the form $\mathcal{S}^q(\mathfrak{Z})$, where $\mathfrak{Z}$ is a finite-dimensional, boundary-less family of links of constant local dimension, and $q \geq 3$. In particular, if X is a polyhedron of constant local dimension r and bdy X is collared in X, then $\mathcal{S}^q(\mathfrak{Z}(X))$ satisfies the hypotheses of Theorem 5.7; and if (rE, rV) is an r-universal pair, then Mor[X, rV] classifies equivalence classes of pairs (M, X) such that M is a manifold whose components are all of dimension

$q + r$, $\mathrm{bdy}\, M \cap X = \mathrm{bdy}\, X$ and $(\mathrm{bdy}\, M, \mathrm{bdy}\, X) < \mathrm{bdy}(M, X)$.

If $\mathfrak{F}$ is not of constant local dimension, we can take $\mathfrak{s} = \mathfrak{s}^{\leq q}(\mathfrak{F})$ for each $q \geq 3$, defined to consist of all pairs (S, F) such that $3 \leq \dim S - \dim F \leq q$.

Remarks 1. The theorem can be generalized to classify, for each X and variety filtration (X) of X, equivalence classes of pairs (M, X) over (X) in which M is a manifold. This would allow (in codimension 2) for local knotting of X in M. It seems to me that not every variety of X can occur as the intrinsic variety of X in some manifold; also that given some variety $\{\mathfrak{X}\}$ of X which does so occur, then the possible types of local knot along the different $\mathfrak{X}_i$ must satisfy some non-trivial relations. If I am correct, then this generalization is of limited value.

2. Using Remark 2 after the Corollary to Proposition 5.5, one can classify equivalence classes of pairs (M, X) over the intrinsic filtration of X such that M is an orientable manifold.

3. Another use of that remark is in dealing with "orientable" polyhedra. A polyhedron X is orientable if: its n-th integral homology group, $H_n(X) = Z$; for every $x \in X$, $H_n(\mathrm{st}(x, X), \ell k(x, X)) = Z$; and the natural map $H_n(X) \to H_n(\mathrm{st}(x, X), \ell k(x, X))$ is an isomorphism for all x in X.

A choice of generator of $H_n(X)$ is an orientation of X. In terms of an n-stratification $\underline{K}$ of X, these conditions can be expressed:

for each block bundle $K_{n,i}/K_{i,i}$, its fibre F is orientable;

the block bundle $K_{n,i}$ has as structural Δ-group the Δ-subgroup

$SP\ell^{\sim}(F)$ of $P\ell^{\sim}(F)$ consisting of those $f : \Delta^a \times F \to \Delta^a \times F$ such that the induced homology isomorphism takes either orientation of F into itself (and not its negative).

Thus for fixed orientable polyhedron X, one can classify the equivalence classes of pairs (M, X) over the intrinsic filtration of X in which M is an orientable manifold.

3. By using a slightly different definition of a "family of link pairs" (namely, replacing link pairs 1 by "link pairs 1: (G, F) is not a suspension-pair") and a slightly different notion of "boundary-less" (namely, allowing a single point to be "boundary-less"), and by using the skeletal filtration of X (instead of its intrinsic filtration), one could similarly classify equivalence classes of pairs (M, X) such that M is a manifold, X is of local codimension at least 3 in M, and $X \subseteq \text{int}\, M$. However, for the applications to cobordism theory which we shall now give, the use of intrinsic filtrations, and not skeletal filtrations, is appropriate.

Let $\mathfrak{F}$ be a finite-dimensional, boundary-less family of links of constant local dimension. Let X, Y be compact $\mathfrak{F}$ polyhedra with bdy $X = \emptyset =$ bdy Y. Then X and Y are $\mathfrak{F}$-cobordant if there exists a compact $\mathfrak{F}$-polyhedron W such that bdy W is the disjoint union $X \cup Y$.

If also V is an $\mathfrak{F}$-cobordism between Y and some Z, then we can form T from the disjoint union $W \cup V$ by identifying the two copies of Y. Then T is an $\mathfrak{F}$-cobordism between X and Z. It follows that

$\mathfrak{J}$-cobordism is an equivalence relation.

Note that if X is $\mathfrak{J}$-cobordant to Y, then $\dim X = \dim Y$. The set of equivalence classes of compact n-dimensional $\mathfrak{J}$-polyhedra is denoted $\eta^n(\mathfrak{J})$. Under the operation of forming the disjoint union, $\eta^n(\mathfrak{J})$ becomes a Z_2-module; the identity element being the class of $X \cup X$ (which bounds $X \times I$), for any boundary-less, n-dimensional $\mathfrak{J}$-polyhedron X.

Let $r\underline{Y}^q/r\underline{U}^q$ be an r-universal $\mathfrak{S}^q(\mathfrak{J})$-stratification. Define the <u>Thom space</u> $Mr\mathfrak{S}^q(\mathfrak{J})$, or Mr^q for short, to be formed from the disjoint union $|r\underline{Y}^q| \cup (\text{fr } r\underline{Y}^q) \times I$ by identifying $\text{fr } r\underline{Y}^q$ to $(\text{fr } r\underline{Y}^q) \times 0$ and $(\text{fr } r\underline{Y}^q) \times 1$ to a point, ∞. Mr^q is a CW-complex (and not, in general, a polyhedron), but $|r\underline{U}^q|$ has closed polyhedral neighbourhoods in Mr^q in which $|r\underline{U}^q|$ has the stratification $r\underline{Y}^q/r\underline{U}^q$ with respect to its intrinsic filtration.

<u>Theorem 5.8.</u> There is a natural isomorphism $\eta^n(\mathfrak{J}) \approx \pi_{n+q}(Mr\mathfrak{S}^q(\mathfrak{J}))$ as abelian groups, whenever $q \geq n+2$ and $r \geq n+1$ (here π_ℓ denotes the ℓth homotopy group).

Given a compact, boundary-less, n-dimensional $\mathfrak{J}$-polyhedron X, there is an embedding $i : X \hookrightarrow S^{n+q}$, which is unique up to isotopy of S^{n+q} since $q \geq n+2$. Let $\underline{\xi}/\underline{K}$ be a stratification of $i(X)$ in S^{n+q} with respect to its intrinsic filtration. Let $g : \underline{K} \mapsto r\underline{U}^q$ classify $\underline{\xi}$. Take a structure $(D^u, B, (r\underline{U}^q \times B)'', \underline{K}'', g'')$ for g. Set $\underline{U}'' = (r\underline{U}^q \times B)''$, and let $\underline{Y}''/\underline{U}''$ be a subdivision of $r\underline{Y}^q \times B$. Then g''^{-1} extends to a p.ℓ. isomorphism $g^* : (|\underline{\xi}|, \text{fr}\,\underline{\xi}) \to (|\underline{Y}'' \restriction \underline{K}''|, \text{fr}\,\underline{Y}'' \restriction \underline{K}'') \subseteq$ $\subseteq (|\underline{Y}''|, \text{fr}\,\underline{Y}'')$. Since $\text{fr}\,\underline{Y}'' = \text{fr}\,r\underline{Y}^q \times D^u$ can be contracted to a point in $Mr^q \times D^u$, g^* extends to a map $f : S^{n+q} \to Mr^q \times D^u$ in which $S^{n+q} - |\underline{\xi}| \to (\text{fr}\,r\underline{Y}^q) \times I \times D^u \subset Mr^q \times D^u$. Let $p : Mr^q \times D^u \to Mr^q$ be the projection; then we define a function $\psi : \eta^n(\mathfrak{J}) \to \pi_{n+q}(Mr^q)$ by mapping the class of X to the class of $p\circ f$.

Conversely, given a map $f : S^{n+q} \to Mr^q$, we can find an f' homotopic to f such that $f'^{-1}(|r\underline{Y}^q| \cup (\text{fr}\,r\underline{Y}^q) \times [0, \frac{1}{2}])$ is a p.ℓ. manifold $P^{n+q} \subseteq S^{n+q}$, with $\text{bdy}\,P = f'^{-1}((\text{fr}\,r\underline{Y}^q) \times \frac{1}{2})$ and f' p.ℓ. on P. We think of S^{n+q} as naturally embedded in a disk D^u, u large. Define a map $f^* : S^{n+q} \to Mr^q \times D^u$ by $f^*(x) = (f'(x), x)$; then $f^* : P \hookrightarrow (|r\underline{Y}^q| \cup (\text{fr}\,r\underline{Y}^q) \times [0, \frac{1}{2}]) \times D^u$ is a p.ℓ. embedding. By Theorem 4.1 there is an isotopy of f^* rel bdy P to some $f^\wedge$ such that $f^\wedge(P) \perp r\underline{Y}^q \times B$, where B is a cell complex triangulation of D^u. Since block transversality implies polyhedral transversality, it follows that X, defined to be $f^{\wedge -1}(|r\underline{U}^q \times B|)$ is an n-dimensional, boundary-less $\mathfrak{J}$-polyhedron. We define a function $\chi : \pi_{n+q}(Mr^q) \to \eta^n(\mathfrak{J})$ by mapping the class of f into the class of X.

The proof that ψ and χ are well-defined, homomorphisms and inverse is the same as Thom's original proof [27, section IV]. One needs the full force of our Addendum 2 to Theorem 4.1 to prove that χ is well-defined.

Examples. 1. Fix n, and take $\mathfrak{J}$ to consist of all boundary-less polyhedra of constant local dimension $\leq n$. Then $\eta^q(\mathfrak{J}) = 0$ for $q \leq n$.

What is $\eta^q(\mathfrak{J})$ for $q \geq n+1$?

2. Let $\mathfrak{J}$ be a finite-dimensional, boundary-less family of links, such that every F in $\mathfrak{J}$ is orientable (in the sense of Remark 3 following Theorem 5.7). Then one can define "oriented" $\mathfrak{J}$-theory. On the other hand, one can take an $\mathfrak{S}^q(\mathfrak{J})$-stratification which is r-universal for orientable $\mathfrak{J}$-polyhedra. Let its Thom space be $MSr\mathfrak{S}(\mathfrak{J})$. Then one can prove the orientable version of our theorem: $\Omega^n(\mathfrak{J}) \approx \pi_{n+q}(MSr\mathfrak{S}^q(\mathfrak{J}))$ whenever $q \geq n+2$ and $r \geq n+1$.

3. Let $\mathfrak{J}$ consist of all joins of complex projective space $CP^2 * CP^2 * \ldots * CP^2$ (including CP^2 and $\emptyset$). Consider an orientable $\mathfrak{J}$-theory in which all the block bundles in Proposition 5.5 are required to be trivial. The analogous construction in the differentiable category gives simply integral homology theory, as Sullivan has pointed out. What is this p.ℓ. theory?

4. Let k be a fixed integer. Let $\mathfrak{J} = \{\emptyset$, the set with k points$\}$. Again require block bundles always to be trivial. Then we have a p.ℓ. version of what Sullivan has called Z_k-theory.

Here is a definition (unpublished) suggested by Akin: Polyhedra X and Y are <u>cobordant</u> if there exists a polyhedron W such that the disjoint union $X \cup Y <$ bdy W, and every component of bdy W meets $X \cup Y$, and such that for every point $w \in W$ there exist points $x \in X$, $y \in Y$ and a p.ℓ. embedding $h : I \hookrightarrow W$ such that:

$w \in \operatorname{im} h$;

$h(0) = x$ and $h^{-1}(X) = 0$;

$h(1) = y$ and $h^{-1}(Y) = 1$;

the intrinsic dimension $d(h(t), W)$ is constant as t varies in $[0, 1]$.

Note that the intrinsic 1-variety of W consists of arcs running from X to Y. Hence on 0-varieties, W induces a bijection $X_o^F = \{x \in X_o : \ell k(x, X) = F\} \longleftrightarrow \{y \in Y_o : \ell k(y, Y) = F\} = Y_o^F$, for each compact F.

Theorem 5.9. Let X and Y be compact, boundary-less polyhedra of constant local dimension. Then there exists a compact cobordism between them if and only if $\mathfrak{F}(X) = \mathfrak{F}(Y)$, X and Y are $\mathfrak{F}(X)$-cobordant, and (the finite sets) X_o^F and Y_o^F have the same number of points, for all F.

Let W be a compact cobordism between X and Y. X_o and Y_o are finite since X and Y are compact, and we have just seen that W induces a bijection $X_o^F \longleftrightarrow Y_o^F$, for all F. Since $\operatorname{bdy} X = \operatorname{bdy} Y = \emptyset$, X and Y are unions of components of bdy W; hence $\operatorname{bdy} W = X \cup Y$. Pick any point $x'' \in X$. Then there is an embedded arc $h[0, 1] \subset W$ which contains x'', with $h(1) \in Y$ and with $h^{-1}(X) = 0$. Hence $h(0) = x''$. Let N be a regular neighbourhood of $h[0, 1]$ in W. Since the intrinsic filtration of $h[0, 1]$ in W is just $h[0, 1]$, we have that $(N, h[0, 1]) \underset{p.\ell.}{\approx} (M, h(0)) \times I$ for some M. And since $X < \operatorname{bdy} W$, $M \underset{p.\ell.}{\approx} st(x'', X)$. Hence $st(h(1), Y) \underset{p.\ell.}{\approx} st(x'', X)$; so $b.\ell k(x'', X) \underset{p.\ell.}{\approx} b.\ell k(h(1), Y)$. Since x'' can be chosen arbitrarily in X, we have $\mathfrak{F}(X) \subseteq \mathfrak{F}(Y)$. By symmetry, $\mathfrak{F}(Y) \subseteq \mathfrak{F}(X)$. A similar argument shows that W is an $\mathfrak{F}(X)$-polyhedron.

Thus W is an $\mathfrak{F}(X)$-cobordism between X and Y.

Conversely, if $\mathfrak{F}(X) = \mathfrak{F}(Y)$ and W is an $\mathfrak{F}(X)$-cobordism between X and Y, then for each $F \in \mathfrak{F}(X)$ let $\mathcal{V}^F = \{w \in W : \text{b.}\ell\text{k.}(w, W) = F\}$. Then $\mathcal{V}^F$ is a sub-polyspace of W, and a manifold of constant local dimension $m = \dim X - \dim F - 1$. We must distinguish the cases $m = 1$ and $m \geq 2$ (note that m cannot be 0). If $m \geq 2$, $\mathcal{V}^F$ must have a component $\tilde{\mathcal{V}}^F$ which meets X, since $F \in \mathfrak{F}(X)$. For each component $\mathcal{V}_s^F$ of $\mathcal{V}^F$ different from $\tilde{\mathcal{V}}^F$, pick points $\tilde{x}_s^F \in \operatorname{int} \tilde{\mathcal{V}}^F$ and $x_s^F \in \operatorname{int} \mathcal{V}_s^F$. Then $\ell k(\tilde{x}_s^F, W) \underset{p.\ell.}{\approx} S^{m-1} * F$ (the join), and $\ell k(x_s^F, W) \underset{p.\ell.}{\approx} S^{m-1} * F$. If $m = 1$, we must choose points $\tilde{x}_s^F$ and x_s^F more carefully: $\mathcal{V}^F$ consists of circles, which are disjoint from $X \cup Y$, and arcs, which meet $X \cup Y$ in their end-points. There is at least one arc $\tilde{\mathcal{V}}^F$ which meets X. List the circles as $\mathcal{V}_1^F, \ldots, \mathcal{V}_r^F$, and for each $s = 1, \ldots, r$ pick points $\tilde{x}_s^F \in \operatorname{int} \tilde{\mathcal{V}}^F$, $x_s^F \in \operatorname{int} \mathcal{V}_s^F$. Some arcs of $\mathcal{V}^F$ run from X to Y. Since there is a bijection $X_o^F \longleftrightarrow Y_o^F$, the other arcs can be listed as: $\tilde{\mathcal{V}}_{r+1}^F, \ldots, \tilde{\mathcal{V}}_t^F$; $\mathcal{V}_{r+1}^F, \ldots, \mathcal{V}_t^F$, where the end-points of $\tilde{\mathcal{V}}_s^F$ are in X, those of $\mathcal{V}_s^F$ in Y, for $s = r+1, \ldots, t$. (It is irrelevant whether $\tilde{\mathcal{V}}^F$ is some $\tilde{\mathcal{V}}_s^F$ or no.) Pick points $\tilde{x}_s^F \in \operatorname{int} \tilde{\mathcal{V}}_s^F$, $x_s^F \in \operatorname{int} \mathcal{V}_s^F$, for $s = r+1, \ldots, t$. As before, we have $\ell k(\tilde{x}_s^F, W) \underset{p.\ell.}{\approx} S^0 * F \underset{p.\ell.}{\approx} \ell k(x_s^F, W)$.

Now for each pair F, s, where $F \in \mathfrak{F}(X)$ and s is in the range described for that particular F, take a copy of $(S^{m-1} * F) \times I$. Let W'' be formed from the disjoint union

$cl[W - \cup\{st(z, W) : z \text{ is some } \tilde{x}_s^F \text{ or } x_s^F\}]$

$\cup \cup\{(S^{m-1} * F) \times I)\}$, by identifying, for each pair F, s, $\ell k(\tilde{x}_s^F, W)$ with $(S^{m-1} * F) \times 0$ and $\ell k(x_s^F, W)$ with $(S^{m-1} * F) \times 1$. (See diagram 9. p. 5.32). We have essentially done a large number of "0-dimensional surgeries" on the intrinsic varieties of W.) Then W'' is an $\mathfrak{F}(X)$-polyhedron with boundary $X \cup Y$. For each $F \in \mathfrak{F}(X)$, let $\mathcal{V}''^F$ be defined analogously to $\mathcal{V}^F$. In the case $m \geq 2$, $\mathcal{V}''^F$ is the connected sum of the components of $\mathcal{V}^F$. Since $\mathcal{V}''^F$ is connected and meets both X and Y, Akin's condition is satisfies for points $w \in \mathcal{V}''^F$. In the case m = 1, we have replaced $\tilde{\mathcal{V}}_s^F \cup \mathcal{V}_s^F$ by a pair of arcs running from X to Y, for each s = r+1, ..., t. We have simultaneously taken the connected sum of one of these arcs with all the circles $\mathcal{V}_s^F$, s = 1, ..., r. Thus $\mathcal{V}''^F$ consists only of arcs running from X to Y, so Akin's condition is satisfied for points $w \in \mathcal{V}''^F$ here too. So W'' is a cobordism between X and Y.

The proof of Theorem 5.8 uses only Addendum 2 to Theorem 4.1. Here is an application of Addendum 1 of that theorem:

Let X be a polyhedron, $(X_n, \ldots, X_0)$ its intrinsic filtration. Let Man(X) be the set of equivalence classes of pairs (M, X) over (X) such that M is a manifold, (X) is the intrinsic filtration of X in M, and X is of local codimension ≥ 3 in M. Now let K be a cell complex triangulation of X; then each X_i is covered by a subcomplex $K_i < K$ for i = 0, ..., n. Let ξ/K be a cS-block bundle. Then there is a map

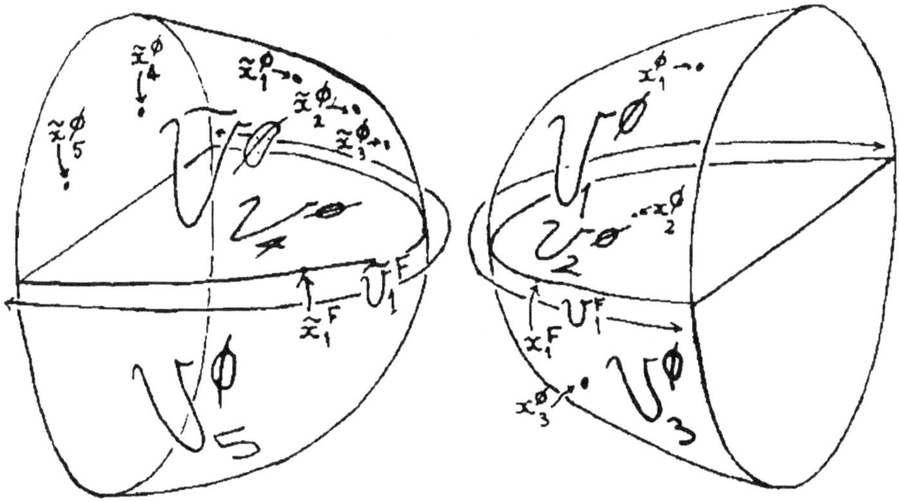

X = "Θ" An $\mathcal{F}(X)$ - cobordism between X and Y Y = "Θ"

$\mathcal{F}(X) = \{F, \emptyset\}$, where F = three points

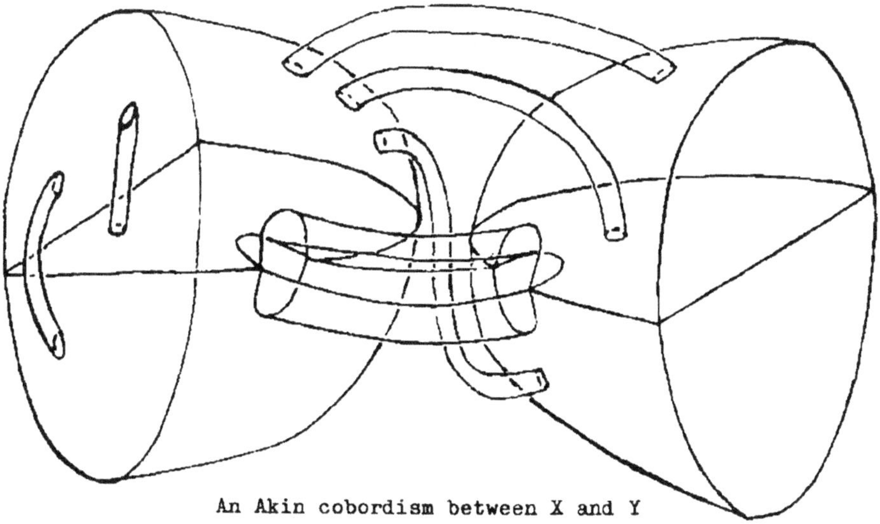

An Akin cobordism between X and Y

diagram 9

$\xi\oplus : \mathrm{Man}(X) \to \mathrm{Man}(|\xi|)$ defined thus: let (M, X) be a pair, then up to equivalence we may assume M is an abstract regular neighbourhood of X, so X is a deformation retract of M. Let L be a cell complex triangulation of M in which X is covered by a subcomplex L'; we may assume L' subdivides K. Take a subdivision ξ'/L' of ξ; then ξ' extends to a cS-block bundle η/L. $|\eta|$ is a manifold, and the equivalence class of $(|\eta|, M)$ is defined to be $\xi\oplus(M, X)$. I leave the reader to check that $\xi\oplus$ is well-defined on the equivalence class of (M, X).

Theorem 5.10. Given X and ξ as above. Then $\xi\oplus$ is a bijection.

Remarks. 1. ξ is not in general uniquely determined by the pair $(|\xi|, X)$; see Chapter 3, p.3.2. For different ξ, $\xi^{\sim}$ giving the same pair $(|\xi|, X)$, the bijections $\xi\oplus$ and $\xi^{\sim}\oplus$ could, nonetheless, be the same. I do not know whether this is always the case or not.

2. The theorem can easily be generalized: Let (X) be a variety filtration of X which refines its intrinsic filtration, and define $\mathrm{Man}'(X)$ to be the set of equivalence classes of pairs (M, X) over (X) such that M is a manifold. Then any ξ/K defines a variety filtration of $|\xi|$ by $(|\xi \restriction K\langle X_n\rangle|, \ldots, |\xi \restriction K\langle X_0\rangle|)$; and ξ induces a map $\xi\oplus : \mathrm{Man}'(X) \to \mathrm{Man}'(|\xi|)$. The proof of the theorem shows that $\xi\oplus$ is a bijection in this situation too.

The proof consists of constructing an inverse function $\varphi : \mathrm{Man}(|\xi|) \to \mathrm{Man}(X)$: Say the fibre of ξ is cS^r and the dimension of

X is k. Let γ/U be a (k+r)-universal cS^r-block bundle. We may assume that $|U|$ is a large-dimensional manifold; then so is $|\gamma|$. Given an abstract regular neighbourhood N of $|\xi|$ such that N is a manifold, let $d : N \to |\xi|$ be a deformation retraction. We have a map $f : X \to |U|$ which classifies ξ; by general position we may assume f is a p. ℓ. embedding. By subdividing ξ and γ if necessary, we may assume $f : K \to U$ is cellular. Then f extends to a block bundle map $f^* : \xi \to \gamma$, which is also an embedding. By general position again, the map $f^* \circ d : N \to |\gamma|$ is homotopic, keeping $|\xi|$ fixed, to an embedding $g : N \hookrightarrow |\gamma|$. Let $\gamma^\sim/U$ be a strict minidivision of γ (in the sense of Rourke and Sanderson's [19, II]; $\gamma^\sim$ is obtained by what I described as "peeling off the frontier" of γ in Chapter 4, p.4.3.) Then the intrinsic filtration of $|\xi|$ is strongly transverse to $\gamma^\sim$. By Addendum 1 to Theorem 4.1 there is an isotopy g_t of g, fixed on $|\xi|$, such that $g_1 N \perp \gamma^\sim$. Set $M = g_1^{-1}(|U|)$. Then $\varphi(N, |\xi|)$ is defined to be (M, X).

One has to show that the equivalence class of (M, X) is independent of choices of d, f, g and g_t. Details are left to the reader; the plan is to work with $N \times I$ and $\gamma \times I/U \times I$. We have choices d, f, g and g_t, giving rise to M, to use with $N \times 0$ and $\gamma \times 0$; and choices d', f', g' and g_t', giving rise to M', to use with $N \times 1$ and $\gamma \times 1$. Now $d \cup d'$ extends over $N \times I$ for homotopy-theoretic reasons. $f \cup f'$ extends over $N \times I$ since γ is universal and by general position. $g \cup g'$ extends over

$N \times I$ by general position. $g_t \cup g'_t$ extends to an isotopy of $N \times I$ in $|\gamma| \times I$, keeping $|\xi| \times I$ fixed, by a collaring argument. By Addenda 1 and 2 to Theorem 4.1, there is a further isotopy of $N \times I$, fixed on $N \times 0$, $N \times 1$ and $|\xi| \times I$, which carries $N \times I$ transverse to $\gamma^{\sim} \times I$. This gives us a pair $(M'', X \times I)$ which extends $(M, X) \times 0$ and $(M', X) \times 1$. By Theorem 3.10, $(M'', X \times I)$ is equivalent to $(M, X) \times I$; hence (M, X) is equivalent to (M', X).

It is now straightforward to check that φ is well-defined on the equivalence class of $(N, |\xi|)$, and that φ is indeed inverse to $\xi \oplus$. This completes the proof of Theorem 5.10.

Chapter 6. Obstructions to Block Bundles

There is an urgent problem left from Chapter 5: when is a p.ℓ. map $f: X \to Y$ stratifiable? This immediately reduces to the case that X is a sub-polyhedron of Y, and $\dim Y - \dim X$ is large (since we replace Y by $Y \times D$, D a large-dimensional disk). Then a necessary condition is that X is locally flat in Y; that means, for every $x \in X$ there is a p.ℓ. isomorphism $f : st(x; Y, X) \to st(x, X) \times (D, v)$ such that $f(x) = x \times v$, where D is a disk, $v \in \text{int } D$. We show that this condition is not sufficient.

We shall be at the same time discussing another question: given $X \subset Y$, when does X have a cS-block bundle neighbourhood in Y? It is again necessary that X be locally flat in Y, and again this condition is insufficient. It is tempting to conjecture that these questions are equivalent; that is, the inclusion $X \hookrightarrow Y$ is stratifiable if and only if X has a cS-block bundle neighbourhood in Y; we allow cS-block bundles to have different dimensions of fibres over different components of the base space throughout this chapter. (This conjecture is proved in Chapter 8). In the present chapter we shall see that if X is locally flat in Y, then the "primary" obstructions to finding a cS-block bundle neighbourhood of X in Y, and to exhibiting the inclusion $X \hookrightarrow Y$ as a stratified map, are the same. But there are definitely further obstructions to both problems.

Thus the results of this chapter should be viewed, not as a description of stratifiable maps, but as a discussion of why block bundles in the sense of Rourke and Sanderson are not sufficient to describe regular neighbourhoods even in the case of a locally flat subpolyhedron of a larger polyhedron. We are thus generalizing Stone [24]; our methods of proof generalize straight-forwardly those of Rourke and Sanderson's [21].

A p.ℓ. map $f : X \to Y$ is locally flat if for some large-dimensional Euclidean space $\mathbb{R}^u$ there are a locally flat sub-polyhedron $X'' \subseteq Y \times \mathbb{R}^u$ and a p.ℓ. isomorphism $g : X \to X''$ such that $f = p \circ g$, where $p : Y \times \mathbb{R}^u \to Y$ is the projection.

If $f : X \to Y$ is locally flat, then $f^{-1}(\mathrm{bdy}^\circ Y) = \mathrm{bdy}^\circ X$. Hence $f^{-1}(\mathrm{bdy}\, Y) = \mathrm{bdy}\, X$. If $h : Y \to Z$ is also locally flat, then $h_\circ f : X \to Z$ is locally flat.

A homotopy $f_t : X \to Y$ is locally flat if it is locally flat regarded as a map $f_t : X \times I \to Y$. A map $f : X \to Y$ is a locally flat homotopy equivalence if there exist a map $g : Y \to X$ and homotopies a_t of X, b_t of Y such that a_o and b_o are the identity of X and Y; $a_1 = g \circ f$, $b_1 = f \circ g$; and f, g, a_t, b_t are all locally flat.

For the rest of this chapter, when we write: $X \subseteq Y$, we shall assume that X is locally flat in Y and that Y is a weak regular neighbourhood of X. Let $\{\mathfrak{X}_n, \ldots, \mathfrak{X}_o\}$ be a variety of X equivalent to its intrinsic variety and such that every $\mathfrak{X}_i$ is connected. Let $\{\mathfrak{Y}_m, \ldots, \mathfrak{Y}_o\}$ be a similar variety of Y. Then our conventions imply that each $\mathfrak{X}_i$ is a submanifold of some $\mathfrak{Y}_j$, and conversely that each $\mathfrak{Y}_j$ has some $\mathfrak{X}_i$ as submanifold. Hence $m = n$, and we may re-number the $\mathfrak{Y}_j$ so that $\mathfrak{X}_i$ is a submanifold of $\mathfrak{Y}_i$, for $i = 0, \ldots, n$. This we shall henceforth assume done; the phrase "intrinsic variety" applied to X or Y will mean the appropriate one of these varieties, and "intrinsic filtration" will mean the appropriate associated filtration.

Now let 'B be a cell complex triangulation of Y in which X is covered by a subcomplex 'A. By Theorem 1.6 there are subcomplexes $'B_i < 'B$ covering Y_i, for $i = 0, \ldots, n$; and then $'A_i = 'A \cap 'B_i$ covers X_i. Let B and B' be first- and second-derived subdivisions of 'B, and let A and A' be the induced subdivisions of 'A. Using these subdivisions, the construction of Theorem 2.1 gives n-stratifications $\underline{K}$ of X dual to 'A, and $\underline{L}$ of Y dual to 'B. We shall refer to $\underline{K}$ and $\underline{L}$ as _simultaneously dual_ to ('B, 'A).

The proof of Theorem 2.1 shows that for each $i = 0, \ldots, n$ we have a decomposition $((i\eta)) = (i\xi, (i\mathfrak{f}))/K_{i,i}$, where ξ is a cS-block bundle and $(i\mathfrak{f})$ an (n-i)-flag, defined thus: Recall that $K_{i,i} = \{s*A_i \text{ and } s*\text{bdy } A_i : s \in A_i \text{ and } \hat{s} \in \mathfrak{X}_i\}$. Then $i\xi(s*A_i \text{ or } s*\text{bdy } A_i) = s*B_i \text{ or } s*\text{bdy } B_i$. (In fact, $i\xi$ is just a normal block bundle for the submanifold $|K_{i,i}|$ of the manifold $|L_{i,i}|$ constructed according to Rourke and Sanderson's [19, I, §4].) $(i\mathfrak{f})$ is just the flag $(K_{n,i}, \ldots, K_{i+1,i})/K_{i,i}$; so $i\eta_{0,j} = K_{i+j,i}$ for $j = 0, \ldots, n-i$. Finally $i\eta_{1,j}(s*A_i \text{ or } s*\text{bdy } A_i) = s*(B_{i+j} \text{ or } s*\text{bdy } B_{i+j})$.

In sum, we see that X has a normal (n+1, n)-ns $\underline{N}$ in Y such that for each $i = 0, \ldots, n$, $(N_{n+1,i}, \ldots, N_{i,i})$ can be given the structure of a decomposition $((i\eta)) = (i\xi; (i\zeta))$, where $(|i\zeta_{n-i}|, \ldots, |i\zeta_o|) = (N_{n,i}, \ldots, N_{i,i})$, and $|i\xi|$ is the ith intrinsic variety of $N_{n+1,i}$. By the uniqueness of normal rns's and Theorem I, 4.A of the Appendix we see that the equivalence class of each $((i\eta))$ (in the sense of the Appendix) depends only on X and Y.

Lemma 6.1(n). Given $X \subset Y$ (with the conventions stated before), say the intrinsic filtration of Y is $Y_n, \ldots, Y_o$. Assume that either: 1. X has a cS-block bundle neighbourhood in Y, or: 2. (Y) has an n-stratification in which X is covered by a restriction. Let $\underline{N}$ be a normal $(n+1, n)$-ns for X in Y, giving rise to decompositions $((i\eta))$, for $i = 0, \ldots, n$. Then each $((i\eta))$ must be a block decomposition.

We use induction on n. Lemma 6.1(0) is trivial. Now assume Lemma 6.1(n-1). First consider the case that X has a cS-block bundle neighbourhood ξ/K in Y. Let $(N'_{n,o}, \ldots, N'_{o,o})$ be a regular neighbourhood of X_o in $(X_n, \ldots, X_o)$. By subdividing ξ/K if necessary, we may assume each $N'_{j,o}$ is covered by a subcomplex of K. Let $N'_{n+1,0} = |\xi \restriction K<N'_{n,o}>|$; then $(N'_{n+1,0}, \ldots, N'_{0,0})$ is a regular neighbourhood of X_o in (Y, X). We can give $(N'_{n+1,0}, \ldots, N'_{0,0})$ the structure of a decomposition $((o\eta'))$ which has $\xi \restriction K<N'_{n,o}>$ as a blocking. Hence $((o\eta'))$ is a block decomposition; and it follows that $((o\eta))$ is a block decomposition, since we can isotop $\underline{N}$ till $(N_{n+1,0}, \ldots, N_{0,0}) = (N'_{n+1,0}, \ldots, N'_{0,0})$. Now set $Y'' = cl[Y - N'_{n+1,0}]$, $X'' = X \cap Y''$. Then ξ restricts to a normal cS-block bundle ξ''/K'' of

X'' in Y'', and we have isotoped $\underline{N}$ till $\underline{N}' = \{N_{j,i} : i \geq 1\}$ is a normal rns for X'' in Y''. So by inductive hypothesis, each $((i\eta))$ is a block decomposition, for $i = 1, \ldots, n$.

In case $X \subseteq Y$ is covered by an n-stratification pair $\underline{K} < \underline{L}$, let $o\xi/oK$ be a normal cS-block bundle neighbourhood for the submanifold X_o of the manifold Y_o. Subdivide $\underline{L}$ till $|o\xi|$ is covered by a subcomplex of $L_{o,o}$; then $(L_{n,o}, \ldots, L_{1,o}) \vdash L_{o,o}\langle o\xi\rangle$ is a blocking of $((o\eta'))$ for suitable choice of $((o\eta'))$. Now the argument continues as before.

<u>Remark.</u> The converse of each statement is false; that is, if all the $((i\eta))$ are block decompositions, X nonetheless need not have a cS-block bundle neighbourhood in Y, nor need $X \subseteq Y$ be covered by a pair $\underline{K} < \underline{L}$. We shall offer an example which is a counterexample to both converse statements, but since it relies on the theory of decompositions which are not block decompositions, we postpone the example to the end of the chapter.

<u>Lemma 6.2.</u> Given $X \subseteq Y$, with the usual conventions. Then there is a locally flat deformation retraction of Y onto X.

Let $('B, 'A)$ be a simplicial triangulation of (Y, X); and let $\underline{K}$ and $\underline{L}$ be n-stratifications of X and Y simultaneously dual to $('B, 'A)$, giving rise to decompositions $((i\eta))$ for $i = 0, \ldots, n$. It suffices to construct a locally flat deformation retraction $a_t : \cup\{|((i\eta))| : i = 0, \ldots, n\} \to$ $\to X$, since $cl[Y - \cup\{|((i\eta))|\}]$ is just a collar. We use induction on a standard ordering of the cells of $\underline{K}$, the inductive step being to extend a_t over $|((i\eta))(\sigma)|$, where $\sigma \in K_{i,i'}$ say. Since $i\xi(\sigma)$ has fibre a disk, there is a p.ℓ. isomorphism of pairs

$p : (|((i\eta))(\sigma)|, |(i\xi)(\sigma)|) \to |(i\xi)(\sigma)| \times (D, v) = |K_{n,i}(\sigma)| \times (D, v)$,

where D is a disk, $v \in \operatorname{int} D$. We may assume that p corresponds $|((i\eta))(\partial\sigma)| \longleftrightarrow |K_{n,i}(\partial\sigma)| \times D$ and $L^{\bullet}_{n,i} \cap (|i\xi(\sigma)|) \longleftrightarrow K^{\bullet}_{n,i}(\sigma) \times D$.
A deformation retraction of D onto v induces a locally-flat deformation retraction b_t of $W = |((i\eta))(\sigma)|$ onto $V = |K_{n,i}(\sigma)|$. Now a_t is already defined on $Z = |((i\eta))(\partial\sigma)| \cup L^{\bullet}_{n,i} \cap (|i\xi(\sigma)|)$. Set $V* = K^{\bullet}_{n,i}(\sigma)$, and let $W*$ be a regular neighbourhood of Z rel V in W. (See diagram 10, p. 6.7). Then $W*$ has the form $Z \times_{V*} I$, with Z corresponding to $Z \times 0$. Define a homotopy of W rel V by:

$c_t(x) = x$, for $x \notin W*$;

$c_t(z, t') = (z, (2t'-t)/(2-t))$, for $t' \geq t/2$;

$= a_{t-2t'}(z, 0)$, for $t' \leq t/2$.

Define $d_{t,t''}(x) = c_{t(1-2t'')}(x)$, for $0 \leq t'' \leq \frac{1}{2}$;

$= b_{t(2t''-1)}(x)$, for $\frac{1}{2} \leq t'' \leq 1$,

for all $x \in W$. Finally define the required extension of a_t by:

$a_t(x) = b_t(x)$, for $x \notin W*$;

$a_t(z, t') = d_{t,t'}(z, t')$, for $(z, t') \in W*$.

(These maps are not piecewise linear as defined, but by using piecewise linear approximations to the functions of t, t' and t'' that we used, a_t becomes piecewise linear.) Then a_t is a locally flat deformation

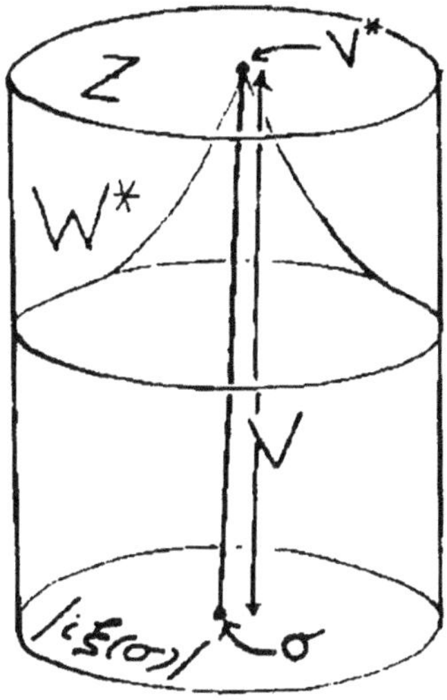

case 1

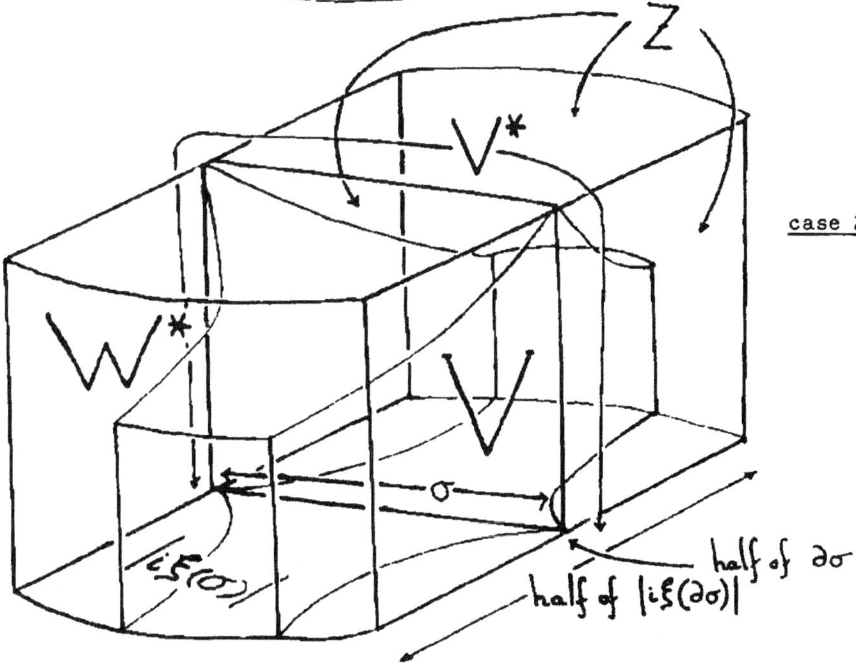

case 2

diagram 10

retraction of W onto V which extends the given a_t on Z.

This completes the inductive step. At the end of the induction we have a deformation retraction a_t of $\cup\{|((i\eta))|\}$ onto $|\underline{K}|$ which restricts, for every $\sigma \in K_{i,i'}$, to a locally flat deformation retraction of $|((i\eta))(\sigma)|$ onto $|K_{n,i}(\sigma)|$. It follows that a_t is locally flat.

Proposition 6.3(n). Given a locally flat homotopy equivalence $f : X \to Y$ such that $f : \text{bdy } X \hookrightarrow \text{bdy } Y$ is an embedding. Assume that for every $x \in X$, $\dim st(fx, Y) \geq \dim st(x, X) + 3$. Then there is a locally flat homotopy h_t of f such that $h_1 : X \hookrightarrow Y$ is an embedding, and h_t equals f on bdy X for all t.

Let $(X_n, \ldots, X_o)$ be the intrinsic filtration of X. We use induction on n. For $n = 0$, this is just the Casson-Sullivan Embedding Theorem (see Casson [5], Haefliger [8], and Rourke [18]). Now assume Proposition 6.3(n-1):

Let $\circ f : X_o \to Y_o$ be the restriction of f. By the Casson-Sullivan theorem there is a homotopy $\circ a_t$ of $\circ f$ rel $\circ f \restriction \text{bdy } X_o$ to an embedding $\circ a_1 : X_o \hookrightarrow Y_o$. Since $\circ a_t$ preserves intrinsic dimension, it follows from Akin [1, Theorem VIII. 2] that $\circ a_t$ extends to a locally flat homotopy a_t of f rel $f \restriction \text{bdy } X$. Henceforth we suppress a_t and assume that $f : X \to Y$ is an embedding on X_o.

Now let $F : X \hookrightarrow Y \times \mathbb{R}^u$ be a locally flat embedding, where $\mathbb{R}^u$ is a large-dimensional Euclidean space, such that $f = p \circ F$, p being the projection $p : Y \times D \to Y$. Let $'A$, $'B$ and $'C$ be simplicial triangulations of X, $Y \times D$ and Y such that F and p are simplicial. Using first- and second-derived subdivisions A, B, C and A', B', C' of 'A, 'B, 'C, we construct n-stratifications $\underline{K}$ of X, $\underline{L}$ of Y x D and $\underline{J}$ of Y dual to 'A, 'B and 'C. To each cell $\sigma \in K_{o,o}$ we associate cells $F^{\wedge}(\sigma) \in L_{o,o}$, $f^{\wedge}(\sigma) \in J_{o,o}$ by the rules: according as $\sigma = s*A_o$ or $s*\text{bdy } A_o$, so $F^{\wedge}(\sigma) = (Fs)*B_o$ or $(Fs)*\text{bdy } B_o$; and $f^{\wedge}(\sigma) = (fs)*C_o$ or $(fs)*\text{bdy } C_o$. Since $f : X_o \to Y_o$ is 1-1, it follows that as maps of sets, $F^{\wedge} : K_{o,o} \to L_{o,o}$ and $f^{\wedge} : K_{o,o} \to J_{o,o}$ are 1-1.

Take a standard ordering of the cells of $K_{o,o}$, and a standard ordering of the cells of $L_{o,o}$ such that if ρ precedes σ in $K_{o,o}$, then $F^{\wedge}(\rho)$ precedes $F^{\wedge}(\sigma)$ in $L_{o,o}$. Assume inductively that we have found a locally flat isotopy F_t of F such that:

whenever ρ precedes σ, $f_1 = p \circ F_1 : |K_{n,o}(\rho)| \hookrightarrow |J_{n,o}(f^{\wedge}\rho)|$ is 1-1 (it is automatically locally flat); $F_t|L_{n,o}(\tau)| = |L_{n,o}(\tau)|$ for every $\tau \in L_{o,o}$ and for all t. To simplify notation we shall suppress F_t, and assume that F already satisfies the first condition. If $\sigma \in \text{bdy } K_{o,o}$, then $|K_{n,o}(\sigma)| \subset \text{bdy } X$, so there is nothing to do. Now assume $\sigma \notin \text{bdy } K_{o,o}$.

As we have seen, $(L_{,o})/L_{o,o}$ and $F((K_{,o})/K_{o,o})$ define a decomposition $((o\eta))/F(K_{o,o}) = (o\xi; (o\zeta))/F(K_{o,o})$. Since $o\xi$ is a cS-block bundle, there is a p. ℓ. isomorphism

$b : (|L_{n,o}(F^{\wedge}(\sigma))|, F(|K_{n,o}(\sigma)|)) = (|((o\eta))(F(\sigma))|, |(o\zeta)(F(\sigma))|) \to$

$\to |K_{n,o}(\sigma)| \times (D', v')$, where D' is a disk, $v' \in \text{int } D'$, which corresponds $|L_{n,o}(F^{\wedge}(\partial\sigma))| \longleftrightarrow |K_{n,o}(\partial\sigma)| \times D'$, and

$L^{\bullet}_{n,o}(F^{\wedge}(\sigma)) \longleftrightarrow K^{\bullet}_{n,o}(\sigma) \times D'$. Hence $F : K^{\bullet}_{n,o}(\sigma) \hookrightarrow L^{\bullet}_{n,o}(F^{\wedge}(\sigma))$ is a locally flat homotopy equivalence.

For $s \in A_o$, the simplicial maps $s \xrightarrow[F]{} Fs \xrightarrow[p]{} fs$ are 1-1. If σ is of the form $s*A_o$, we take dual cells to F_s and fs in B and C (recall that $|B| = |C| \times \mathbb{R}^u$) to obtain $(Fs)*B \underset{p.\ell.}{\approx} (fs)*C \times D$, where D is a u-disk; similarly if σ is of the form $s*\text{bdy } A_o$, we use $(Fs)*\text{bdy } B \underset{p.\ell.}{\approx} (fs)*\text{bdy } C \times D$. This gives a p.$\ell$. isomorphism c such that the diagram

$$\begin{array}{ccc}
c : |L_{n,o}(F^{\wedge}(\sigma))| & \longrightarrow & |J_{n,o}(f^{\wedge}(\sigma))| \times D \\
 & & \big\downarrow \text{projection} \\
\bigcap\big| & & |J_{n,o}(f^{\wedge}(\sigma))| \\
 & & \bigcap| \\
Y \times \mathbb{R} & \xrightarrow[\text{projection}]{} & Y
\end{array}$$

commutes.

It follows from the natural way in which c is defined that c corresponds $|L_{n,o}(F^{\wedge}(\partial\,\sigma))| \longleftrightarrow |J_{n,o}(f^{\wedge}(\sigma))| \times D$; and hence $L^{\bullet}_{n,o}(F^{\wedge}(\sigma)) \longleftrightarrow J^{\bullet}_{n,o}(f^{\wedge}(\sigma)) \times D$. Thus the projection $p : Y \times \mathbb{R} \to Y$ restricts to a locally flat homotopy equivalence $p : L^{\bullet}_{n,o}(F^{\wedge}(\sigma)) \to J^{\bullet}_{n,o}(f^{\wedge}(\sigma))$. Combining with F shows that $f = p \circ F$ restricts to a locally flat homotopy equivalence

$d : K^{\bullet}_{n,o}(\sigma) \to J^{\bullet}_{n,o}(f^{\wedge}(\sigma))$.

By inductive hypothesis, d is 1-1 on $K^{\bullet}_{n,o}(\partial\sigma)$. This need not be all of bdy $K^{\bullet}_{n,o}(\sigma)$; but if G is the fibre of $K_{n,o}(\sigma)$, then bdy $K^{\bullet}_{n,o}(\sigma) = K^{\bullet}_{n,o}(\partial\sigma) \cup$ bdy $G \times \sigma$ (using some trivialization of $K_{n,o}(\sigma)$). And bdy $G \times \sigma \subset$ bdy X, where f is 1-1. So d is 1-1 on all of bdy $K^{\bullet}_{n,o}(\sigma)$. By Proposition 6.3(n-1) there is a locally flat homotopy d_t of d rel bdy $K^{\bullet}_{n,o}(\sigma)$ such that $d_1 : K^{\bullet}_{n,o}(\sigma) \hookrightarrow J^{\bullet}_{n,o}(\sigma)$ is an embedding. Since the disk D is large-dimensional, we can find an isotopy e_t of $F \restriction K^{\bullet}_{n,o}(\sigma)$ in $L^{\bullet}_{n,o}(F^{\wedge}(\sigma))$ rel bdy $K^{\bullet}_{n,o}(\sigma)$ such that $d_t = p\circ e_t$ for all t; it follows that e_t is locally flat. By Theorem 1.6, e_t extends to an isotopy g_t of $L^{\bullet}_{n,o}(F^{\wedge}(\sigma))$. Since bdy $L^{\bullet}_{n,o}(F^{\wedge}(\sigma)) \cap F(K^{\bullet}_{n,o}(\sigma)) = F(\text{bdy } K^{\bullet}_{n,o}(\sigma))$, we can choose g_t to be the identity on bdy $L^{\bullet}_{n,o}(F^{\wedge}(\sigma))$. Extend g_t over $|\underline{L}|$ by the standard procedure. Then $p\circ g_1\circ F : |K_{n,o}(\sigma)| \hookrightarrow |J_{n,o}(f^{\wedge}(\sigma))|$ is just the conical extension of $f \restriction K_{n,o}(\partial\sigma)$ and d_1, and so is an embedding. If ρ precedes σ, then g_t is the identity on $|L_{n,o}(F^{\wedge}(\rho))|$, so $p\circ g_1\circ F$ is still the embedding f on $|K_{n,o}(\rho)|$. Finally g_t is locally flat (trivially). And we can choose g_t to be the identity on bdy $|\underline{L}|$. So $F_t = g_t\circ F$ is the required isotopy. This completes the inductive step in the induction on the standard ordering of $K_{o,o}$.

At the end of the induction we have a locally flat homotopy of f rel bdy X to some f' such that

$f' : |K_{n,o}| \hookrightarrow |J_{n,o}|$ is an embedding. Now set

$X'' = |\mathcal{C}_o(\underline{K})|$, $Y'' = |\mathcal{C}_o(\underline{L})|$, $f'' = f' \restriction X''$. Then bdy $X'' = (X'' \cap \text{bdy } X) \cup \cup \dot{K}_{n,o}$. The hypotheses of Proposition 6.3(n-1) are thus satisfied; so there is a locally flat homotopy h''_t of f'' rel bdy X'' to an embedding. Then h_1, defined to be f' on $|K_{n,o}|$ and h''_1 on $|\mathcal{C}_o(\underline{K})|$ is the required embedding. This completes the inductive step in proving Proposition 6.3(n). Hence Proposition 6.3 holds for all n.

Lemma 6.4. Given a decomposition $((\eta)) = (\xi;\zeta)/K$, where ξ is a cS-block bundle, and ζ a cF-block bundle. Then there is a cF-block bundle ζ'/K with $|\zeta'| \subseteq |((\eta))|$ such that the cbb's ξ, ζ' and $\eta_{1,1}$ form the block decomposition $[\xi;\zeta']$. ζ' is called complementary to ξ in $((\eta))$. Further if ζ'' is another complement to ξ in $((\eta))$, then there is an isotopy f_t of $|((\eta))|$ such that $f_1 : \zeta' \approx \zeta''$; f_t is the identity on $|\xi|$; and for every $\sigma \in K$, $f_t|((\eta))(\sigma)| = |((\eta))(\sigma)|$ for all t.

The proof is essentially the same as that of Rourke and Sanderson's [19, II, Theorem 5.1], and uses the fact that since $|\xi|$ is a manifold (recall from Chapter 4 our convention according to which $|K|$ is a manifold), $|((\eta))|$ has the structure of a cF-block bundle $\nu/|\xi|$. Then ζ' is just $\nu \restriction |K|$.

Let ξ/K and η/L be cF-block bundles. A p.ℓ. map $f : |\xi| \to |\eta|$ is blockwise locally flat if for every $\sigma \in K$ there is a $\tau \in L$ such that $f|\xi(\sigma)| \subseteq |\eta(\tau)|$ and $f : |\xi(\sigma)| \to |\eta(\tau)|$ is locally flat, with $f(\sigma) \subseteq \tau$ and $f(\dot{\xi}(\sigma)) \subseteq \dot{\eta}(\tau)$.

If $|K|$ and $|L|$ are manifolds, it follows that f itself is locally flat. Similarly one defines blockwise locally flat homotopies and homotopy equivalences between cF-block bundles.

For any compact polyhedron F we can now define the Δ-set $G^{\sim}(F)$ to have for k-simplexes any blockwise locally flat homotopy self-equivalence of the trivial cF-block bundle, $f : \Delta^k \times cF/\Delta^k \to \Delta^k \times cF/\Delta^k$, such that f is the identity on $|\Delta^k| = |\Delta^k| \times c$. (See Rourke and Sanderson's [22, I] for the definition of a Δ-set.) The Δ-set $P\ell^{\sim}(F)$ is defined to be the set of f in $G^{\sim}(F)$ such that f is a p.ℓ. isomorphism. (In the notation of Chapter 5, $P\ell^{\sim}(F)$ would have been called $P\ell^{\sim}(F, F)$.) Then $G^{\sim}(F)$ is a Δ-monoid, and $P\ell^{\sim}(F)$ is a Δ-subgroup. Let $G/P\ell^{\sim}(F)$ be the Δ-set of right cosets. $G/P\ell^{\sim}(F)$ can be geometrically realized as a CW-complex [22, I], which we again call $G/P\ell^{\sim}(F)$.

Let $BP\ell^{\sim}(F)$ be the base space of a universal $P\ell^{\sim}(F)$-block bundle, constructed according to [22, II, Corollary 2.3]. Let $BG^{\sim}(F)$ be the base space of a universal $G^{\sim}(F)$-block bundle, according to [22, II, Proposition 3.10]. As in [22, II, (3.18)] one can show that there is (up to homotopy type) a fibration $\pi : BP\ell^{\sim}(F) \to BG^{\sim}(F)$ whose fibre is $G/P\ell^{\sim}(F)$. Let $p : BP\ell^{\sim}(S^q) \times BP\ell^{\sim}(F) \to BP\ell^{\sim}(F)$ be the projection, and let $\lambda : E \to BP\ell^{\sim}(S^q) \times BP\ell^{\sim}(F)$ be the fibration with fibre $G/P\ell^{\sim}(F)$ induced from π by $\pi \circ p$. Thus λ is covered by a map of fibrations $\lambda^{-} : E \to BP\ell^{\sim}(F)$. E can be represented as $\{w, x, y\} \in BP\ell^{\sim}(S^q) \times BP\ell^{\sim}(F) \times BP\ell^{\sim}(F) : \pi x = \pi y\}$. λ thus has the section s defined by $s(w, x) = (w, x, x)$.

Theorem 6.5. Assume $q \geq 2$. Then E is a classifying space for $(cS^q- ; cF-)$ decompositions. That is, for every cell complex K, let Dec(K) be the set of isomorphism classes of decompositions $(\xi; \zeta)/K$ with ξ a cS^q-block bundle, ζ a cF-block bundle; then there is a functorially defined bijection $\theta : [|K|, E] \longleftrightarrow \mathrm{Dec}(K)$.

The proof is essentially the same as the argument of Rourke and Sanderson's [21, section 3], so we omit some details of the present proof.

Let $((\eta)) = (\xi; \zeta)/K$ be a decomposition, and let ζ'/K be complementary to ξ in $((\eta))$. Then we have maps $f : |K| \to BP\ell^{\sim}(S^q)$ classifying ξ, and $g, g' : |K| \to BP\ell^{\sim}(F)$ classifying ζ and ζ' respectively. Let $(V, \{V(\sigma)\})$ be a regular neighbourhood of $\dot{\xi}$ in $(\dot{\eta}_{1,1}, \{\dot{\eta}_{1,1}(\sigma)\})$ such that $V \cap \dot{\zeta} = \emptyset$ and $V \cap \dot{\zeta}' = \emptyset$. Set $W = \mathrm{cl}[\dot{\eta}_{1,1} - V]$, $W(\sigma) = W \cap \dot{\eta}_{1,1}(\sigma)$, for every $\sigma \in K$. Then $(W, \{W(\sigma)\})$ is a weak regular neighbourhood of $\dot{\zeta}$ and of $\dot{\zeta}'$. (For $W(\sigma)$ is a weak regular neighbourhood of $\dot{\zeta}(\sigma)$ and of $\dot{\zeta}'(\sigma)$ since $(\xi; \zeta)$ and $[\xi; \zeta']$ are both trivial over σ; and one fits the $W(\sigma)$ together by induction on a standard ordering of K, using Proposition 1.4 to ensure one always has a weak regular neighbourhood.) By Lemma 6.2 the inclusions $\dot{\zeta} \hookrightarrow W$ and $\dot{\zeta}' \hookrightarrow W$ are blockwise locally flat homotopy equivalences. Hence we have a blockwise locally flat homotopy equivalence $h : \dot{\zeta} \to \dot{\zeta}'$. That is, ζ and ζ' are isomorphic as $G^{\sim}(F)$-bundles, and so $\pi \circ g = (\pi \circ p) \circ (f \times g')$ (up to homotopy). Thus associated to $((\eta))$ we have a map $\gamma : |K| \to E$ such that $\lambda \circ \gamma = f \times g'$, $\bar{\lambda} \circ \gamma = g$ (up to homotopy).

This process defines a function φ : Dec(K) $\to$ $[|K|, E]$.

Conversely, given a map $\gamma : |K| \to E$, we have a cS^q-block bundle ξ/K classified by $p_1 \circ \lambda \circ \gamma$, and a cF-block bundle ζ'/K classified by $p \circ \lambda \circ \gamma$; where p_1 and p are the projections of $BP\tilde{\ell}(S^q) \times BP\tilde{\ell}(F)$ onto its first and second factors. We also have a cF-block bundle ζ/K classified by $\pi \circ \lambda^{-} \circ \gamma$; and ζ is block-wise locally flat homotopy equivalent to ζ'. Set $((\eta')) = [\xi ; \zeta']$. Then the previous argument shows that we have a blockwise locally flat homotopy equivalence $h : \zeta^{\bullet} \to W' \subset \eta'^{\bullet}_{1,1}$, where W' is defined as before. The argument of [21] shows how to find a blockwise locally flat homotopy of h to an embedding $h'' : \zeta^{\bullet} \hookrightarrow W'$: one uses induction on a standard ordering of K, and if $h : \zeta^{\bullet}(\partial\sigma) \hookrightarrow W'(\partial\sigma)$ is already an embedding, then by a double application of Proposition 6.3 one can homotop h on $\zeta^{\bullet}(\sigma)$ rel $\zeta^{\bullet}(\partial\sigma)$ to an embedding, first on bdy $\zeta^{\bullet}(\sigma)$, and then on $\zeta^{\bullet}(\sigma)$. By the standard procedure, h'' extends to a blockwise locally flat embedding $h' : |\zeta| \hookrightarrow |((\eta'))|$. Then we have a decomposition $((\eta))/K$ defined by the block bundles ξ, $h'(\zeta)$, $\eta'_{1,1}$. This process defines the map $\theta : [|K|, E] \to \mathrm{Dec}(K)$.

The proof that θ and φ are well-defined and inverse is the same as in [21].

Remark. The statement of the theorem is imprecise because our polyhedra have to be finite-dimensional. (The same difficulty occurred in Chapter 5). A precise statement would say that for each integer $r \geq 0$ there is an "approximation" rE to E such that $\varphi : [|K|, E] \to \mathrm{Dec}(K)$ is a bijection whenever $\dim K \leq r$.

Corollary 1. Let ξ/K be a cS^q-block bundle with $q \geq 2$, and ζ/K a cF-block bundle. Let $f : |K| \to BP\mathcal{L}^{\sim}(S^q)$, $g : |K| \to BP\mathcal{L}^{\sim}(F)$ classify ξ and ζ. Then $[\xi ; \zeta]$ is classified by $s \circ (f \times g) : |K| \to E$.

Corollary 2. Let $((\eta)) = (\xi ; \zeta)/K$ be a decomposition, with ξ a cS^q-block bundle, $q \geq 2$, and ζ a cF-block bundle. Then $((\eta))$ is a block decomposition if and only if its classifying map $\gamma : |K| \to E$ is homotopic through sections of λ to $s \circ \lambda \circ \gamma$. The obstructions to doing so lie in the cohomology groups $H^n(|K|, \pi_n G/P\mathcal{L}^{\sim}(F))$.

Corollary 3. Let $((\eta)) = (\xi ; \zeta)/K$ as before, and let ξ'/K be a $cS^{q'}$-block bundle. By extending ξ' over $|((\eta))|$, one obtains a new decomposition $((\xi' \oplus \eta)) = (\xi' \oplus \xi ; \zeta)$. Then $((\eta))$ is a block decomposition if and only if $((\xi' \oplus \eta))$ is.

Theorem 6.6. Given $X \subset Y$ locally flat and of local codimension at least 3. For each $F \in \mathfrak{F}(X)$, let $X^F = \{x \in X : b.\ell k(x, X) = F\}$. Then for each F there is a sequence of obstructions $\alpha_n(F) \in H^n(X^F, \pi_n G/P\mathcal{L}(F)^{\sim})$, for $n = 0, 1, 2, \ldots$ (each defined only if all previous $\alpha_i(F)$ are 0) such that if either: 1. X has a cS-block bundle neighbourhood in Y; or 2. there is an intrinsic n-stratification of Y in which X is covered by a restriction, - then the $\alpha_n(F)$ must be 0 for all n and F. Moreover these obstructions are stable in the sense that if Z has the structure of a cS-block bundle over Y, then the obstructions for X in Y are the same as those for X in Z.

This follows from Corollaries 2 and 3 just given and Lemma 6.1.

In particular, take F to be the disjoint union of S^r and a point. Then $G/P\ell^{\sim}(F)$ is isomorphic to (the standard) $G/P\ell_r = G/P\ell^{\sim}(S^r)$. If $r \geq 2$, the homotopy groups $\pi_n(G/P\ell_r)$ are known to be: 0, if n is odd; $\mathbb{Z}$, if $n \equiv 0(4)$; and $\mathbb{Z}_2$ if $n \equiv 2(4)$ (see Sullivan [25], Kervaire and Milnor [11], or Rourke [18].) Hence whenever $q \geq 2$, $r \geq 2$ and n is even, there are locally flat embeddings of $S^n \times cF$ in $S^n \times cF \times D^{q+1}$ which do not admit block bundle neighbourhoods.

By modifying this example we can show that the obstructions described in Theorem 6.6 are not sufficient to ensure that X has a cS-block bundle in Y: Write $S^n = D^n_+ \cup D^n_-$, as two hemi-spheres, and let v_+ be the centre of D^n_+. We can choose a locally flat embedding $i : S^n \times cF^r \hookrightarrow$ $\hookrightarrow S^n \times cF^r \times D^{q+1}$ so that $i(S \times cF)$ has no normal block bundle, and i restricts to the obvious map $D^n_+ \times cF \to D^n_+ \times cF \times 0 \subset S \times cF \times D^{q+1}$, where $0 \in \operatorname{int} D^{q+1}$. Let Y be formed from the disjoint union $(S \times cF \times D^{q+1}) \cup (D^{q+1} \times I)$ by identifying $v_+ \times c \times D^{q+1}$ to $D^{q+1} \times 0$ in the obvious way. Similarly let X be formed from $i(S \times cF) \cup (0 \times I)$ by identifying $i(v_+ \times c)$ to 0×0. Then X is locally flat in Y and has intrinsic filtration: $X_{n+r+1} = X$, $X_n = i(S^n \times c) \cup I$, $X_1 = I$, and $X_0 = i(v_+ \times c)$. Thus each $X_i - X_{i-1}$ is an open disk and so the $((i\eta))$ must all be trivial, and so block decompositions. Hence the obstructions of

Theorem 6.6 all vanish. However, if X had a normal cS^q-block bundle ξ/K in Y, then one sees by considering intrinsic dimensions that $i(S \times cF)$ would be covered by a subcomplex $K' < K$, and that $|\xi| \cap (S \times cF \times D^{q+1})$ would equal $|\xi \restriction K'|$. Thus $\xi \restriction K'$ would provide a block bundle structure for $i(S \times cF)$ in $S \times cF \times D^{q+1}$, which is a contradiction.

Similarly, if there were a pair of intrinsic $(n+r+1)$-stratifications $\underline{K} < \underline{L}$ covering $X \subseteq Y$, then one sees that

$\{|\beta| \cap (S \times c S^r \times D^{q+1}) : \beta \in L_{,i}$ for some $i \leq n\}$

would form a cS^r-block bundle μ/P over

$P = \{|\gamma| \cap (S \times c \times D^{q+1}) : \gamma \in L_{n+q+1,i}$ for some $i \leq n\}$.

And μ/P would restrict to a cS^r-block bundle neighbourhood μ'/P' of $i(S \times c)$ in $S \times cS^r$ (see diagram 11, p. 6.19). Then there would be a complementary blocking ν/Q to μ/P of $i(S \times cS^r)$ in $S \times cS^r \times D^{q+1}$ which would restrict to a cS^q-bb neighbourhood of $i(S \times c)$ in $S \times c \times D^{q+1}$. But ν/Q could then be extended to a cS^q-bb neighbourhood of $i(S \times cF)$ in $S \times cF \times D^{q+1}$, which is a contradiction.

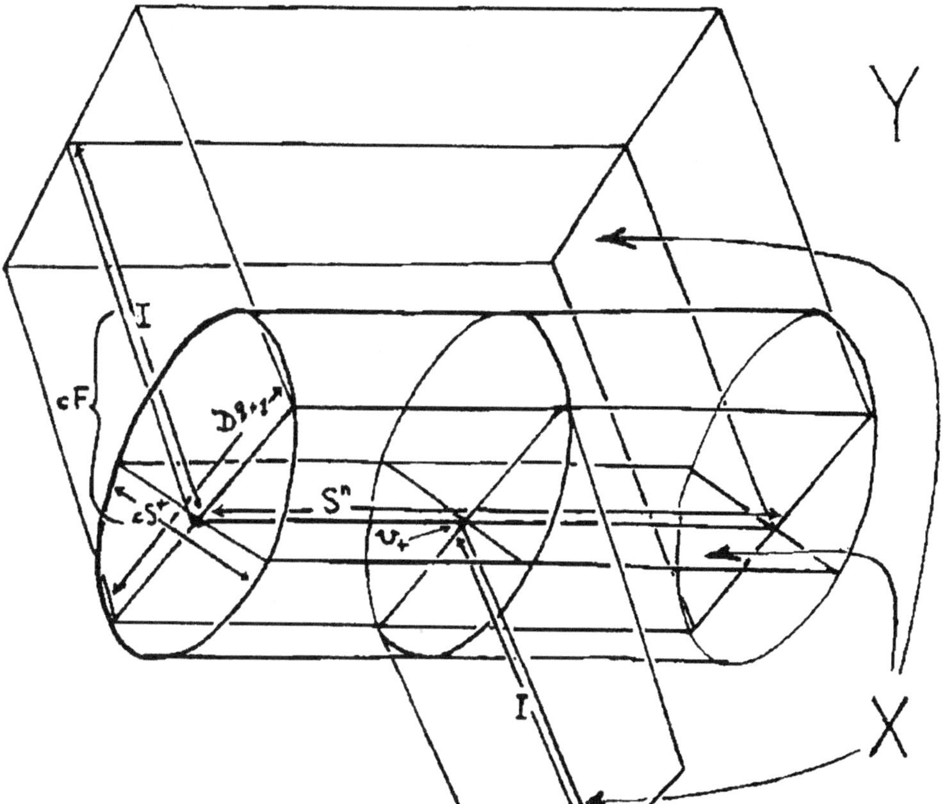

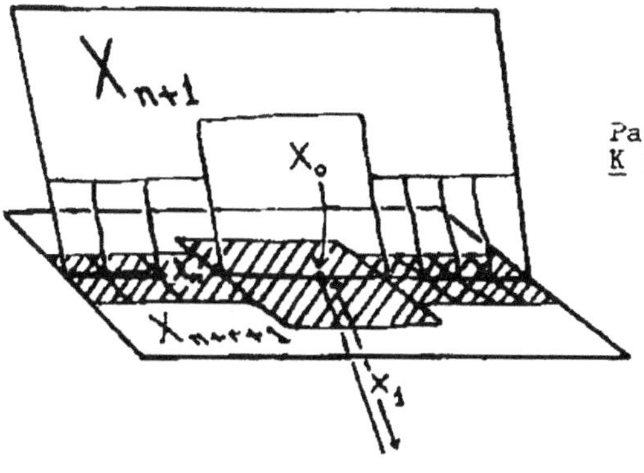

Part of an (n+r+1) - stratification $\underline{K}$ on X. Blocks of μ'/P' are shaded.

<u>diagram 11</u>

Chapter 7. Open Questions

In this chapter I shall discuss some of the unsolved problems and unexplored paths whose investigation appears interesting and profitable.

i. At the root of most of the unsolved problems I have raised in the text is this:

Let $((\eta)) = [\xi ; \zeta]/K$ be a block decomposition, say of two cS-block bundles. Recall how we construct a pair of complementary blockings of $((\eta))$. First we take the block decomposition $((\xi \times \zeta)) = [\xi \times K;\ K \times \zeta]/K \times K$. Subdivide to $((\xi \times \zeta))'/(K \times K)'$ so that the diagonal is covered by a subcomplex Δ_K, and consider the restriction $((\xi \times \zeta))' \restriction \Delta_K$. Note that subdivision of block bundles is not unique, so $((\xi \times \zeta))' \restriction \Delta_K$ is unique only up to some sort of isotopy, as we shall see. We construct a special pair of blockings of $((\xi \times \zeta))' \restriction \Delta_K$: Let B be a cell complex triangulation of $|\xi|$ in which every block of ξ/K is covered by a subcomplex of B; then $B \times \zeta/B \times K$ is a blocking of $((\xi \times \zeta))$. A complementary blocking in this case is constructed by symmetry. To block $((\xi \times \zeta))' \restriction \Delta_K$ we subdivide everything. Let $(\xi \times K)'/(K \times K)'$ be a subdivision of $(\xi \times K)/K \times K$. Let B' be a cell complex subdivision of $B \times K$ such that every block of $(\xi \times K)'/(K \times K)'$ is covered by a subcomplex. Then there is a subcomplex $B'_\Delta < B'$ which covers $|(\xi \times K)' \restriction \Delta_K|$. Subdivide $B \times \zeta/B \times K$ to $(B \times \zeta)'/B'$. Then $(B \times \zeta)' \restriction B'_\Delta$ defines a choice of $((\xi \times \zeta))' \restriction \Delta_K$ together with a blocking. By symmetry, we obtain a complementary blocking of what is, in general, a different choice of $((\xi \times \zeta))' \pitchfork \Delta_K$. We can arrange these choices to be allowably isotopic. Now for the general $((\eta)) = [\xi ; \zeta]$, there is an isomorphism $((\eta)) \approx ((\xi \times \zeta))' \pitchfork \Delta_K$; and pulling back the two blockings of $((\xi \times \zeta))'$ gives up complementary blockings μ and ν of $((\eta))$. The question is: do we have to admit allowable isotopies? or can we choose blockings of the same choice of $((\xi \times \zeta))' \pitchfork \Delta_K$? Can we make the blocks of μ and ν intersect "orthogonally"? Can we get from

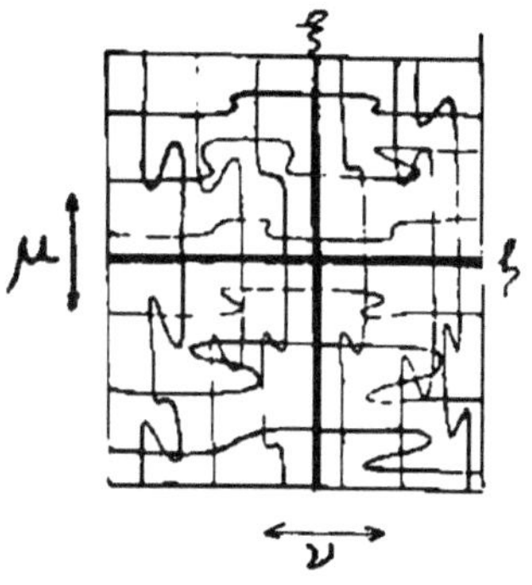

to

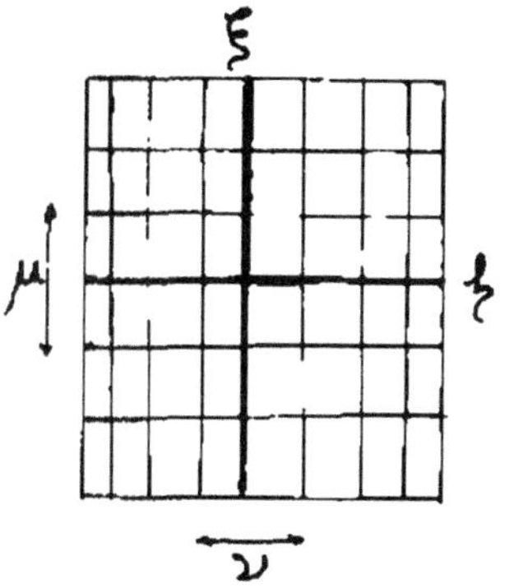

?

I suspect that the answer, in general, is "no".

ii. The main problem is still that left unanswered in Chapters 5 and 6: given $X \subset Y$ locally flat, when is the inclusion $X \to Y$ a stratifiable map? In particular, does this happen if, only if, or if and only if X has a cS-block bundle neighbourhood in Y? (The answer is "if and only if"; see Chapter 8.)

A question that might be more amenable is: In the context of Theorem 5.10, does the bijection $\xi \oplus : \mathrm{Man}(X) \to \mathrm{Man}(|\xi|)$ depend on the choice of ξ, for fixed $(|\xi|, X)$?

iii. With a better definition of a "stratifiable map", problem ii, as stated, may become irrelevant. However, it can be recast in the form: is block transversality between a polyhedron and a manifold in a manifold symmetric? The question: is block transversality in a manifold symmetric? is presumably more difficult, and still remains unanswered.

iv. The previous problems could probably be avoided if one had a canonical way of reducing block bundles to p.ℓ. fibre bundles. This cannot be done in general; see Rourke and Sanderson [20]. Find a class of polyhedra and regular neighbourhoods in which this can be done, and whose elements can be recognized in p.ℓ. topology. For example, give a p.ℓ. characterization of p.ℓ. stratifications that come from Thom's "ensembles stratifiés" [28].

v. Generalize the definition of a "stratifiable" map between polyhedra. For example, give a general class of p.ℓ. maps $f : X \to Y$ such that given an abstract regular neighbourhood of Y one can functorially induce an abstract regular neighbourhood of X. Or else define a p.ℓ. analogue of Thom's "morphismes stratifiés" [28]; in particular, what is a p.ℓ. "morphisme sans éclatement"? Sullivan has suggested that an analogue of "morphismes stratifiés" might be "semi-triangulable" maps; that is, continuous maps $f : X \to Y$ such that the p.ℓ. structures on X and Y extend to a p.ℓ. structure on the mapping cylinder of f. Then (he suggests) "morphismes sans éclatement" would be just the p.ℓ. maps. I should warn the reader that this definition must be slightly modified since, as it stands, the composition of semi-triangulable maps need not be semi-triangulable.

vi. Thom [28] also has a much more general notion of transversality than ours. For example, if V is an "ensemble stratifié", and $\{\mathcal{V}_i\}$ the underlying p.ℓ. variety of V, then the projections $p_1 : V \times \mathbb{R}^1 \to V$ and $p_2 : V \times \mathbb{R}^1 \to \mathbb{R}^1$ are both morphisms in Thom's sense, whereas only p_1 is a stratifiable p.ℓ. map. p_2 is "transverse" in Thom's sense to any point of $\mathbb{R}^1$. (Note, by the way, that any abstract regular neighbourhood of $\mathbb{R}^1$ induces an abstract regular neighbourhood of $V \times \mathbb{R}^1$. The reason we did not allow p_2 as a stratifiable map is that this procedure does not yield manifolds from manifolds in general.) How should p.ℓ. transversality be generalized so that p_2 is transverse to any point of $\mathbb{R}^1$? Another problem in this area is: prove a relative transversality theorem for polyhedra X and Y in a polyhedron M. This would be very interesting even if one assumed special properties of the pairs (M, X) and (M, Y); for example, that X and Y are locally flat in M. One might well want a different notion of transversality. D. Ravenel (unpublished) has shown that for any polyhedron X, there is an isotopy f_t of the diagonal Δ_X in $X \times X$ such that $f_1\Delta_X \cap \Delta_X$ is a finite set of points. Thus $f_1\Delta_X$ is in some sense transverse to Δ_X.

vii. To define the Whitney sum of two (n+1, n)-stratifications over the same n-stratification seems to me difficult and not very useful in general. One case, however, tantalizes me: for manifolds one has the equation: tangent block bundle ⊕ stable normal block bundle = trivial block bundle. For any polyhedron one can define the tangent, stable normal and trivial stratifications. Give geometric meaning to the analogous equation.

viii. The obstructions that arise in dealing with stratified polyhedra are usually of two kinds: those existing in the flags $(K_{,i})/K_{i,i'}$ and those that have to do with the way these flags are fitted together.

The former obstructions are usually straightforward, if incomputable. The latter obstructions are quite mysterious. They might become less so if one had a working definition of a "principal stratification", which would generalize the notion of a principal bundle. A test case would be to describe the higher obstructions to deciding whether X has a cS-block bundle neighbourhood in Y, given that X is locally flat in Y (following up Chapter 6).

ix. Given a polyhedron X, let $(X_n, \ldots, X_o)$ be its intrinsic filtration. A <u>resolution</u> of X is a p.ℓ. surjection $f : M \to X$ such that M is a manifold and f is 1-1 on $f^{-1}(X-X_{n-1})$. Develop an obstruction theory for resolving any given X. As an ideal goal, prove geometrically Hironaka's theorem [10] that any complex algebraic variety can be resolved. Kato and Sullivan have results on the resolution problem. Their method has been to resolve "from the top down". They look for a sequence of partial resolutions $(n-1)f : (n-1)X \to X, \ldots, if : iX \to (i+1)X, \ldots,$ $of : oX \to 1X$ such that for all i, if is a surjection; $iX_{n-1} = \ldots = iX_{i-1}$; and if is 1-1 on $if^{-1}((i+1)X - (i+1)X_i)$. Then oX is a manifold, and the composition $(n-1)f \circ, \ldots, \circ of$ is a resolution of X. My own instinct is to resolve "from the bottom up"; that is, to look for a sequence $o'f : o'X \to X, \ldots, i'f : i'X \to (i-1)'X, \ldots, (n-1)'f : (n-1)'X \to (n-2)'X$ such that for all i, i'f is a surjection; $i'X_i = \emptyset$; and i'f is 1-1 on $i'f^{-1}((i-1)'X - (i-1)'X_i)$. Then (n-1)'X is a manifold, and $o'f \circ \ldots \circ (n-1)'f$ a resolution of X. The advantage is that each i'X is an $\mathfrak{J}(X)$-polyhedron, so one retains some control over the "singular sets" of the i'X - this is not in general the case with the iX. The disadvantage is that one has to know $\mathfrak{J}$ (X)-theory fairly well, which leads one back to problem viii - whereas with the iX one needs only cobordism theory using manifolds ($\mathfrak{J}$ ($\emptyset$)-theory).

x. Zeeman [31] has defined a marvelous spectral sequence on any polyhedron X, which I think will be useful in attacking problems of the sort raised in viii. and ix. McCrory [15] has considerably simplified Zeeman's derivation of the spectral sequence and brought out its intimate connection with the cap product. He has shown its relation to an intrinsic stratification of X. He has also proved Zeeman's conjecture that the "filtration" of a homology class on X equals the "codimension" of that class.

I should also like to see this spectral sequence used to investigate "transversality" between two homology classes in X. My feeling is that "purely" transverse intersection cannot always be obtained, and that there are "homology operations" which describe the "impurely" transverse part of the intersection of the given homology classes. Such "homology operations" might also have to do with problem i., whose context resembles that of Steenrod and Epstein's definition of squaring operations [23].

Chapter 8. Symmetry of Transversality

In this chapter we prove that given $X, Y \subseteq M$ with M a manifold, Y a submanifold of M, then Y is block transverse to X if and only if X is block transverse to Y. Hence we can characterize stratifiable maps: a p.ℓ. map $f : X \to Y$ is stratifiable if and only if there are a large-dimensional disk D and a factorization $f = p \circ g$ such that $p : Y \times D \to Y$ is the projection, $g : X \to Y \times D$ is an embedding, and gX has a cS-block bundle neighbourhood in $Y \times D$.

Our method is based on a more flexible use of block decompositions than that of Chapter 4. Since block decompositions were defined by means of a blocking, one should be able to talk about complementary blockings without mentioning a block decomposition. We pick up the thread from p. 4.8.

A p-flag $(\mu)/P$ and a q-flag $(\nu)/Q$ are <u>complementary</u> if:

<u>comp 1</u> there are subcomplexes $\dot{P} < P$ and $jP < P$ for $j = 0, \ldots, q$, with $qP = P$;

<u>comp 2</u> there is a (p, q)-decomposition $((\eta)) = ((\xi); (\zeta))/K$ such that $|\eta_{i,j}| = |\mu_i \restriction jP|$ and $\dot{\eta}_{i,j} = \dot{\mu}_p \cup |\mu_p \restriction \dot{P}|$;

<u>comp 3</u> there is a subdivision $(\mu')/P'$ of (μ) which is a blocking of $((\eta))$;

<u>comp 4</u> there are subcomplexes $\dot{Q} < Q$ and $iQ < Q$ for $i = 0, \ldots, p$ such that $|\eta_{i,j}| = |\nu_j \restriction iQ|$, $\dot{\mu}_p = |\nu_q \restriction \dot{Q}|$, and $|\mu_p \restriction \dot{P}| = \dot{\nu}_q$;

comp 5 there are a subdivision $(\nu')/Q'$ of (ν) and an allowable isotopy f_t of $((\eta))$ such that $f_1(\nu')$ is a blocking of $((\eta))$ complementary to (μ').

Recall that an admissable isotopy of $((\eta))$ (we shall also say: of (μ), or: of (ν)) is a p.ℓ. isotopy f_t of $(|\eta_{p,q}|, \{|\eta_{i,j}|, (\mu^\bullet), (\nu^\bullet)\})$ rel $|K|$. f_t is allowable if it is the identity on $|(\xi)| \cup |(\zeta)|$. Note that the subcomplexes ${}^\bullet P$ and jP are determined by (μ) and (ν), since $|{}^\bullet P| = \nu^\bullet_q \cap |P|$ and $|jP| = |\nu_j| \cap |P|$; similarly for ${}^\bullet Q$ and iQ. Thus the criteria for f_t to be admissable or allowable can be stated wholly in terms of (μ) or (ν). Also note that given (μ), ${}^\bullet P$ and jP satisfying comp 1 and 2 (which can be checked by looking at $st(x; \{|\mu_i \nmid jP|, |\mu_p \nmid {}^\bullet P|\})$ for all $x \in |\mu_p|$), then there exists a q-flag $(\nu)/Q$ complementary to (ν).

Lemma 8.1. Let $(\mu)/P$ and $(\nu)/Q$ be complementary flags, and let a_t be an admissable isotopy of (μ) rel $|P|$. Then $a_1(\mu)/P$ and $(\nu)/Q$ are complementary.

We may assume that (μ) and (ν) are already complementary blockings of a block decomposition $((\eta)) = [(\xi); (\zeta)]/K$. Now a_t defines a p.ℓ. isomorphism a of $|((\eta))| \times I$ rel $|(\zeta)| \times I$ (called the trace of a_t), namely: $a(x, t) = (a_t(x), t)$. So we have the block decomposition structure $((\eta^\wedge)) = [(\xi^\wedge); (\zeta) \times I]/K \times I$ on $|((\eta))| \times I$ defined by $a(((\eta)) \times I)$. And $((\eta^\wedge))$ has the blocking $(\mu^\wedge)/P \times I$ defined as $a((\mu) \times I)$. Let $Q^\wedge$ be a common cell complex subdivision of $Q \times I$ and

$a(Q \times I)$ in which every block of $a((\xi) \times I)$ and its rim are covered by subcomplexes. Set $Q^* = Q^{\wedge}\langle Q \times 0\rangle$, $Q^{*\prime} = Q^{\wedge}\langle Q \times 1\rangle$. Take subdivisions $(\nu^*)/Q^*$ of $(\nu) \times 0/Q \times 0$, and $(\nu^{*\prime})/Q^{*\prime}$ of $a_1(\nu) \times 1/a_1 Q' \times 1$. Then $(\nu^*) \cup (\nu^{*\prime})$ is a blocking of $((\eta^{\wedge})) \upharpoonright K \times \{0, 1\}$ complementary to $(\mu^{\wedge}) \upharpoonright P \times \{0, 1\}$. Let $(\nu^{\wedge})/Q^{\wedge}$ be a blocking of $((\eta^{\wedge}))$ which extends (ν^*) and $(\nu^{*\prime})$ and is complementary to $(\mu^{\wedge})$. Let $(\nu^{\sim})/Q \times I$ be the amalgamation of $(\nu^{\wedge})$. Then $(\nu^{\sim})$ agrees with $(\nu) \times I$ over $Q \times 0 \cup$ $\partial Q \times I$, since a is the identity there; moreover $|\nu_j^{\sim} \upharpoonright (iQ \times I)| = |\nu_j \upharpoonright (iQ \times I)|$ for all i and j. By the Appendix, Corollary I, 1.8, the identity flag isomorphism $id : (\nu) \times I \upharpoonright (Q \times 0 \cup \partial Q \times I) \approx (\nu^{\sim}) \upharpoonright (Q \times 0 \cup \partial Q \times I)$ extends to a flag isomorphism $b : (\nu) \times I \approx (\nu^{\sim})$ mod $Q \times I$. b is in fact a p.ℓ. isomorphism of $|((\eta))| \times I$. The proof of the Appendix, Corollary I, 1.8 shows that one may choose b to be level-preserving; so we can define an isotopy b_t of $|((\eta))|$ whose trace is b. Then b_t is an allowable isotopy of (μ), and $b_1(\nu)$ has the subdivision $(\nu^{*\prime})$ which is a blocking of $a_1((\eta))$ complementary to $a_1(\mu)$. Hence (ν) is complementary to $a_1(\mu)$.

<u>Corollary 1.</u> Let $(\mu)/P$ and $(\nu)/Q$ be complementary flags. Let c_t be an admissable isotopy of (μ) rel $|\partial P|$. Then $c_1(\mu)$ and (ν) are complementary.

We have only to express c_t as the composition of two admissable isotopies a_t and b_t of (μ) rel $|P|$ and $|Q|$ respectively. Let a_t' be the isotopy of $|P| \cup |Q|$ defined by the identity on $|P|$ and the restriction of c_t on $|Q|$. Let $((\eta))$ be a block decomposition as specified in comp 2. There is a variety on $|((\eta))|$ with terms:

$|\eta_{i,j}| - (|\eta_{i-1,j}| \cup |\eta_{i,j-1}| \cup \dot{\eta}_{i,j})$;

$(\dot{\eta}_{i,j} \cap \dot{\mu}_p) - (\dot{\eta}_{i-1,j} \cup \dot{\eta}_{i,j-1} \cup |\mu_p \curlywedge \dot{P}|)$;

$(\dot{\eta}_{i,j} \cap |\mu_p \curlywedge \dot{P}|) - (\dot{\eta}_{i-1,j} \cup \dot{\eta}_{i,j-1})$,

for $i = 0, \ldots, p$ and $j = 0, \ldots, q$. Now $|P| \cup |Q|$ is a union of manifolds of this variety. By Akin's Covering Isotopy Theorem (our Theorem 1.6), a'_t extends to an isotopy a_t of the whole variety. Then a_t is an admissable isotopy of (μ) rel $|P|$. Set $b_t = c_t \circ a_t^{-1}$; then b_t is an admissable isotopy of (μ) rel $|Q|$. A double application of the lemma proves this corollary.

Corollary 2. If $(\mu)/P$ and $(\nu)/Q$ are complementary, then any subdivisions $('\mu)/'P$ of (μ) and $('\nu)/'Q$ of (ν) are complementary.

We may assume that there are subdivisions $(\mu')/P'$ of (μ) and $(\nu')/Q'$ of (ν) which are complementary blockings of a block decomposition $((\eta))$. By further subdividing (μ') and (ν') we may assume P' and Q' are cell complex subdivisions of $'P$ and $'Q$. Let $('\mu')/P'$ and $('\nu')/Q'$ be subdivisions of $('\mu)$ and $('\nu)$. By the Appendix, Theorem I, 4.1, there are isotopies f_t of $|(\mu)|$ rel $|P|$ and g_t of $|(\nu)|$ rel $|Q|$ such that:

$f_1 : ('\mu') \approx (\mu')$, $g_1 : ('\nu') \approx (\nu')$;

$f_t('\mu')$ and $g_t('\nu')$ are subdivisions of (μ) and (ν) for all t.

Hence f_t and g_t are admissable isotopies of (μ), and $f_1('\mu')$ and

$g_1('\nu')$ are complementary. By Corollary 1, $('\mu')$ and $('\nu')$ are complementary. It follows from the definition that $('\mu)$ and $('\nu)$ are complementary.

Lemma 8.2. Given a p-flag $(\mu)/P$ and complementary q-flags $(\nu)/Q$ and $('\nu)/'Q$ (using the same $^{*}P$ and jP). Then (after subdividing (ν) and $('\nu)$ if necessary) there is an admissable isotopy f_t of (μ) rel $|Q|$ such that $f_1 : (\nu) \approx ('\nu)$.

Let the subdivisions $(\mu')/P'$ of (μ), $(\nu')/Q'$ of (ν') and the block decomposition $((\eta)) = [(\xi);(\zeta)]/K$ be given by comp 2, 3 and 5 for (μ) and (ν); we may assume (ν') is already a blocking of $((\eta))$ complementary to (μ'). Let $('\mu')/'P'$, $('\nu')/'Q'$ and $(('\eta)) = [('\xi);('\zeta)]/'K$ be given similarly for (μ) and $('\nu)$. Let $(\zeta^{\wedge})/K^{\wedge}$ be a q-flag structure on $(|qP|, \dots, |oP|) \times I$ which extends $(\zeta) \times 0$ and $('\zeta) \times 1$. Let $P^{\wedge}$ be a cell complex subdivision of $P \times I$ in which every block of $(\zeta^{\wedge})$ and its rim is covered by a subcomplex; we may choose $P^{\wedge}$ to extend $P' \times 0$ and $'P' \times 1$. Let $(\mu^{\wedge})/P^{\wedge}$ be a subdivision of $(\mu) \times I$ which extends $(\mu') \times 0$ and $('\mu') \times 1$. Then $(\mu^{\wedge})$ defines a block decomposition structure $((\eta^{\wedge})) = [(\xi^{\wedge});(\zeta^{\wedge})]/K^{\wedge}$ on $|(\mu)| \times I$ of which $(\mu^{\wedge})$ is a blocking; and $((\eta^{\wedge}))$ extends $((\eta)) \times 0$ and $(('\eta)) \times 1$. Let $(\nu^{\wedge})/Q^{\wedge}$ be a blocking of $((\eta^{\wedge}))$ complementary to $(\mu^{\wedge})$ which extends $(\nu') \times 0$ and $('\nu') \times 1$. Now $|Q^{\wedge}| = |Q| \times I$, and $|\nu_j^{\wedge} \vdash iQ^{\wedge}| = |\nu_j \vdash iQ| \times I$ for all i, j. The proof of Lemma 8.1 shows that, after subdividing $(\nu^{\wedge})$ and $(\nu) \times I$ if necessary, there is an isomorphism $f : (\nu^{\wedge}) \approx (\nu) \times I$ rel $|Q| \times I$; and we can use f to construct an admissable isotopy f_t of (μ) rel $|Q|$. Then f_t is the required isotopy.

Corollary 1. Let $((\eta)) = [(\xi); (\zeta)]/K$ be a block decomposition, (μ) a blocking of $((\eta))$, $(\nu)/Q$ and $('\nu)/'Q$ blockings of $((\eta))$ complementary to (μ). Then (after subdividing (ν) and $('\nu)$ if necessary) there is an admissable isotopy f_t of (μ) rel $|Q|$ such that $f_1 : (\nu) \approx ('\nu)$, and $f_t |((\eta))(\sigma)| = |((\eta))(\sigma)|$ for every cell σ of K.

The proof is by induction on a standard ordering of the cells of K. One needs a slightly strengthened version of Lemma 8.2 to know that if (ν) and $('\nu)$ agree in $|((\eta))(\partial\sigma)|$, then there is an admissable isotopy of $|((\eta))(\sigma)|$ rel $|(\xi)(\sigma)| \cup |((\eta))(\partial\sigma)|$ which carries $(\nu) \restriction Q\langle(\xi)(\sigma)\rangle$ isomorphically to $('\nu) \restriction 'Q\langle(\xi)(\sigma)\rangle$ (after subdividing if necessary). Details are left to the reader.

Corollary 2. Given a p-flag $(\mu)/P$ and subcomplexes $^{\bullet}P < P$ and $jP < P$ for $j = 0, \ldots, q$, such that comp 1 and 2 are satisfied. Let $P^* < P$ be a subcomplex such that $|P^*| < \mathrm{bdy}(|P|, \{|jP|, |^{\bullet}P|\})$. Let $(\nu^*)/Q^*$ be a q-flag complementary to $(\mu) \restriction P^*$ (using $^{\bullet}P \cap P^*$ and the $jP \cap P^*$). Then (ν^*) extends to a q-flag $(\nu)/Q$ complementary to (μ).

An $(n+1, n)$-stratification ξ/K and a cS-block bundle ν/Q are complementary if:

for each $i = 0, \ldots, n$ and $j = i, \ldots, n+1$ there are subcomplexes

$(j, i)Q < Q$ and $(j, i)^{\bullet}Q < Q$ covering $|\xi_{j,i}| \cap |Q|$ and $\xi^{\bullet}_{j,i} \cap |Q|$;
for each $i = 0, \ldots, n$ there is a subcomplex $\nu K_{i,i} < K_{i,i}$ covering $|\nu| \cap |K_{i,i}|$;
for each i, the flag $(\xi_{,i}, K_{,i}) \nmid \nu K_{i,i}$ and $\nu \nmid (n+1, i)Q$ are complementary. See diagram 12, p. 8.8.

We write $(j, i)\nu$ and $(j, i)^{\bullet}\nu$ for $\nu \nmid (j, i)Q$ and $\nu \nmid (j, i)^{\bullet}Q$. An isotopy f_t of $(|\nu|, |Q|)$ or of $(|\xi_{n+1,}|, |K_{n,}|, \ldots, |K_{o,}|)$ is admissable if it restricts to an admissable isotopy of each $\nu \nmid (n+1, i)Q$. By Lemma 8.1, $f_1\nu$ and $\underline{\xi}$, or ν and $f_1\underline{\xi}$, are still complementary. Also if $\underline{\xi}'$ and ν' are subdivisions of $\underline{\xi}$ and ν, then $\underline{\xi}'$ and ν' are complementary, by Corollary 2 to Lemma 8.1 - note that one applies this corollary independently to each flag $(\xi', K'_{,i}) \nmid \nu K'_{i,i}$, rather than trying to isotop the whole stratification. Observe that if $\underline{\xi}$ and ν are complementary, then $\underline{\xi}$, the various complexes $\nu K_{i,i}$ and their subcomplexes $1(\nu K_{i,i}), 0(\nu K_{i,i})$ and $^{\bullet}(\nu K_{i,i})$ specified in comp 1 suffice to determine $(|\nu|, \{|Q|, \nu^{\bullet}\}) \cap |\xi_{n+1,}|$. Similarly ν and the various complexes $(j, i)Q$ and $(j, i)^{\bullet}Q$ suffice to determine the $(n+1, n)$-ns $|\underline{\xi}| \cap |\nu|$.

Proposition 8.3(n). Given $X, Y \subseteq M$ with M a manifold, Y a submanifold of M. Let $(X_n, \ldots, X_o)$ be a variety filtration of X in M, and $\underline{\xi}/\underline{K}$ a stratification of (X) in M. Assume that $Y \pitchfork \underline{\xi}$. Then Y has a normal cS-block bundle ν/Q in M which is complementary to $\underline{\xi}$; so in particular, $X \perp \nu$. Further, given that $X^* = X \cap \text{bdy } M$

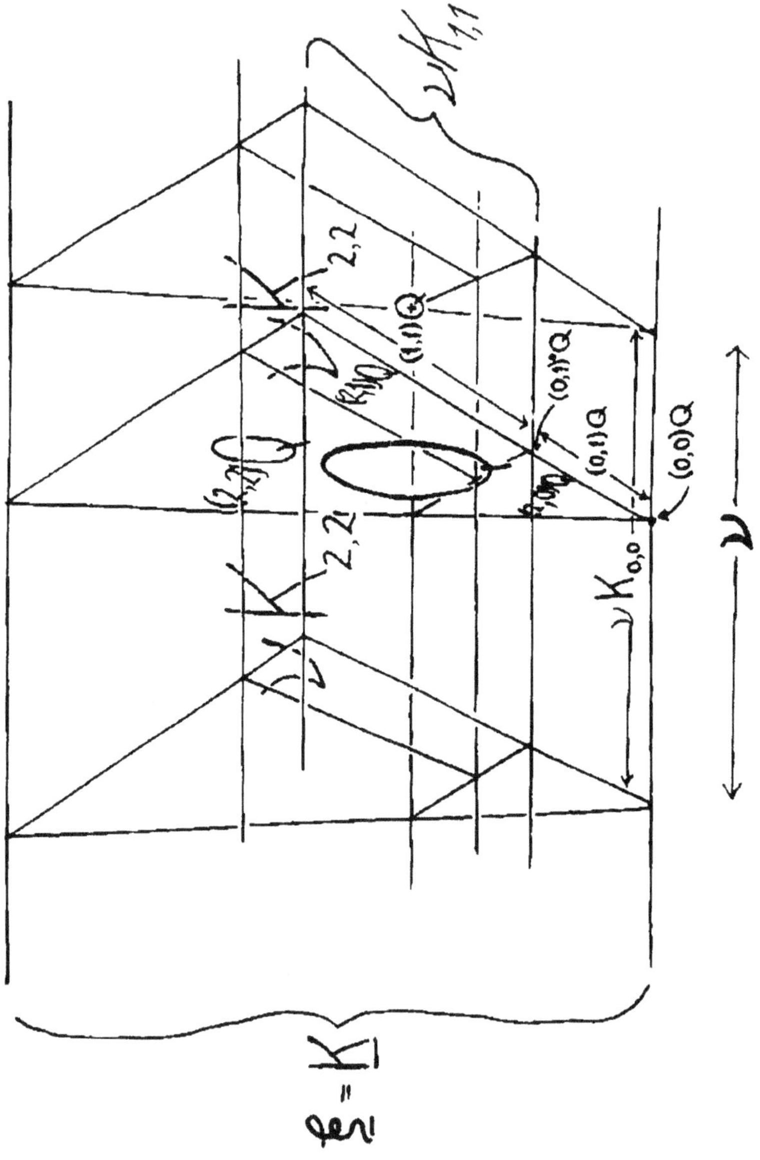

diagram 12

is $< \mathrm{bdy}_M(X)$ and given a normal cS-block bundle ν^*/Q^* of bdy Y in bdy M such that $\underline{\xi}$ respects X^* and ν^* is complementary to $\underline{\xi}^*$; then we may choose ν to extend ν^*.

We use induction on n. Proposition 8.3(0) is Rourke and Sanderson's [19, II, Proposition 4.10].

Now assume Proposition 8.3(n-1):

$Y \cap X_o$ is a submanifold of X_o. Let λ/L be a normal cS-block bundle of $Y \cap X_o$ in X_o; we may assume that λ extends $(0,0)\nu^*$. Subdivide $\underline{\xi}$ so that every block of λ and its rim are covered by subcomplexes of $K_{o,o}$. Then $(\xi, K_{,o}) \upharpoonright K_{o,o}\langle\lambda\rangle$ is a blocking of a block decomposition $((\eta)) = [\lambda ; (\xi, K_{,o}) \upharpoonright K_{o,o}\langle L\rangle]$ over L. To simplify notation I shall henceforth write λK, $(\lambda\sigma)K$ and σK for $K_{o,o}\langle\lambda\rangle$, $K_{o,o}\langle\lambda(\sigma)\rangle$ and $K_{o,o}\langle\sigma\rangle$, where $\sigma \in L$.

We use induction on a standard ordering of the cells of L to construct a blocking μ/P of $((\eta))$ complementary to $(\xi, K_{,o}) \upharpoonright \lambda K$ such that μ is in fact complementary to the weak $(n+1, n)$-stratification $\Lambda(\lambda, \underline{\xi})$. Assume that whenever ρ precedes σ, we have a blocking $\rho\mu/\rho P$ of $((\eta)) \upharpoonright \langle\rho\rangle$ complementary to $\Lambda(\lambda(\rho), \underline{\xi})$ such that $\rho\mu$ extends $\pi\mu$ whenever $\pi < \rho$. The flag $(\xi, K_{,o}) \upharpoonright (\lambda\sigma)K$ is trivial; say its fibre is $(G, F_n, \ldots, F_1)$, and let $h : (\xi, K_{,o}) \upharpoonright (\lambda\sigma)K \to (\lambda\sigma)K \times c(G, F)$ be a trivialization. Let $\underline{\zeta}/\underline{H}$ be a stratification of (F) in G; then $(\lambda\sigma)K \times c(G, F)$ can be completed to a weak $(n+1, n)$-stratification by adding $(\lambda\sigma)K \times \underline{\zeta}$. Thus we have two $(n, n-1)$-stratifications of

$(K_{,o})^{\cdot} \vdash (\lambda\sigma)K$ in $\xi^{\cdot}_{n+1,0} \vdash (\lambda\sigma)K$, namely: $\Lambda^*(\lambda(\sigma), \underline{\xi})$ and $h^{-1}((\lambda\sigma)K \times \zeta)$, both of which define completions of the flag $(\xi, K_{,o}) \vdash (\lambda\sigma)K$. The proof of Lemma 3.2 shows that there is an isotopy a_t of $|(\xi, K_{,o}) \vdash (\lambda\sigma)K|$ rel $|\lambda(\sigma)|$ such that (after subdividing $\underline{\xi}$ and $(\lambda\sigma)K \times \zeta$) $a_1 : h^{-1}((\lambda\sigma)K \times \zeta) \approx \Lambda^*(\lambda(\sigma), \underline{\xi})$ and a_t is an isomorphism of the flag $(\xi, K_{,o}) \vdash (\lambda\sigma)K$ for all t.

Now let (B, A) be a cell complex triangulation of the sequence (G, F) in which every block of ζ and its rim are covered by subcomplexes. Then $(\lambda \vdash \langle\sigma\rangle) \times cB$ is a cS-block bundle structure $\mu^{\wedge}/P^{\wedge}$ on $\sigma \times cG$ in $|\lambda(\sigma)| \times cG$. $\mu^{\wedge}$ is clearly complementary to the stratification $(\lambda\sigma)K \times c\zeta = ((\lambda\sigma)K \times c(G, F)) \cup ((\lambda\sigma)K \times \zeta)$. So $'\mu/'P$, defined to be $a_1 h^{-1}\mu^{\wedge}/a_1 h^{-1}P^{\wedge}$ is complementary to the stratification $\Lambda(\lambda(\sigma), \underline{\xi})$. (This is a trivial consequence of the definitions, since a_t acts on $h^{-1}\mu^{\wedge}$ <u>and</u> on $h^{-1}((\lambda\sigma)K \times c\zeta)$.

Let $\partial'\mu/\partial'P$ be the restriction of $'\mu$ to $\partial'P = {}'P\langle\xi_{n+1,0} \vdash \partial\sigma K\rangle$; and let $\partial\mu/\partial P$ be $\cup\{\rho\mu/\rho P : \rho \lneq \sigma\}$. We shall make $\partial'\mu$ equal $\partial\mu$; then our inductive step in the construction of μ will be complete. Now $\partial\mu$ and $\partial'\mu$ are both complementary to the flag $(\xi, K_{,o}) \vdash (\lambda\partial\sigma)K$. By Lemma 8.2 there is an admissable isotopy b_t of $((\eta)) \vdash \partial\sigma$ rel $|\partial'P|$ such that (after subdividing $\partial\mu$ and $\partial'\mu$ if necessary), $b_1 : \partial'\mu \approx \partial\mu$. We want $b_t\, \partial'\mu$ to be complementary to the stratification $\Lambda(\lambda(\partial\sigma), \underline{\xi})$ for all t. This happens if b_t respects the normal $(n, n-1)$-ns $|\Lambda^*(\lambda(\partial\sigma), \underline{\xi})|$ of $(K_{,o})^{\cdot} \vdash (\lambda\partial\sigma)K$ in

$\xi^{\bullet}_{n+1,0} \vdash (\lambda\, \partial\sigma)K$, by Lemma 8.1, since b_t is the identity on $\xi^{\bullet}_{n+1,0} \vdash (\partial\sigma)K$.

We use the uniqueness theorem for relative regular neighbourhoods quite straightforwardly, except for the cumbersome notation. We work in $\Lambda(\lambda\,(\partial\sigma), \xi) \times I$. The trace of b_t is a p.ℓ. isomorphism b of $|\Lambda\,(\lambda(\partial\sigma), \xi)_{n+1,}| \times I$. Then $b(|\Lambda^*(\lambda(\partial\sigma), \xi)| \times I)$ and $|\Lambda^*(\lambda(\partial\sigma), \xi)| \times I$ are both normal $(n, n-1)$-ns's for $((K_{,o})^{\bullet} \vdash (\lambda\partial\sigma)K) \times I$ in $(\xi^{\bullet}_{n+1,0} \vdash (\lambda\partial\sigma)K) \times I$, which agree in $(\xi^{\bullet}_{n+1,0} \vdash (\lambda\partial\sigma)K) \times \{0, 1\}$ and in $(\xi^{\bullet}_{n+1,0} \vdash (\partial\sigma)K) \times I$ (recall that b_t is the identity on the second term). Assume by induction on i that

$$b(|\Lambda^*(\lambda(\partial\sigma), \xi)_{j,i'}| \times I) = |\Lambda^*(\lambda(\partial\sigma), \xi)_{j,i'}| \times I$$

whenever $1 \leq i' \leq i-1$ and $i' \leq j \leq n+1$. Then the families $(b(|\Lambda^*(\lambda(\partial\sigma), \xi)_{j,i}| \times I) : j = i, \ldots, n+1)$ and $(|\Lambda^*(\lambda(\partial\sigma), \xi)_{,i}| \times I)$ are both regular neighbourhoods of V, which is:

$$\mathrm{cl}[K^{\bullet}_{i,o} \vdash (\lambda\partial\sigma)K - \cup\{|\Lambda^*(\lambda(\partial\sigma), \xi)_{n+1,i'}| : 1 \leq i' \leq i-1\}] \times I$$
$$\cup\ \mathrm{cl}[\xi^{\bullet}_{n+1,0} \vdash (\partial\sigma)K - \cup\{|\Lambda^*(\partial\sigma, \xi)_{n+1,i'}|\}] \times I$$

rel W, which is:

$$\mathrm{cl}[\xi^{\bullet}_{n+1,0} \vdash (\partial\sigma)K - (\cup\{|\Lambda^*(\partial\sigma, \xi)_{n+1,i'}|\} \cup |\Lambda^*(\partial\sigma, \xi)_{n+1,i}|)] \times I$$

in Z, which is:

$$\mathrm{cl}[(\xi, K_{,o})^{\bullet} \vdash (\lambda\, \partial\sigma)K - \cup\{|\Lambda^*(\lambda(\partial\sigma), \xi)_{n+1,i'}|\}] \times I$$

which agree in Z^*, which is:

$$\mathrm{cl}[(\xi, K_{,o})^{\bullet} \vdash (\lambda\, \partial\sigma)K - \cup\{|\Lambda^*(\lambda(\partial\sigma), \xi)_{n+1,i'}|\}] \times \{0, 1\}.$$

(See diagram 13, p. 8.12.) It follows that there is an isotopy d'_t of

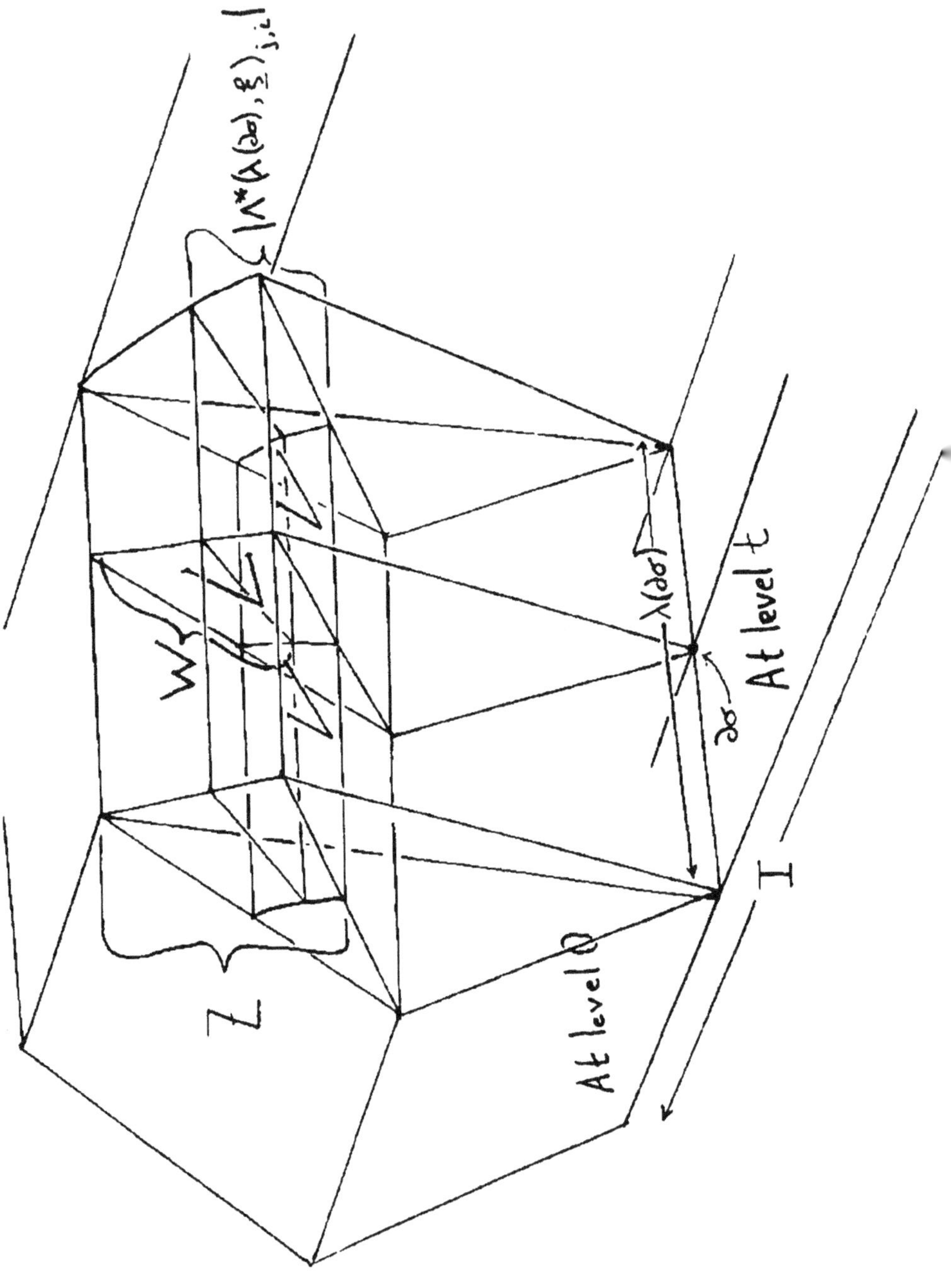

diagram 13

Z rel $Z^* \cup V$ such that the families $(d'_1 b(|\Lambda^*(\lambda(\partial\sigma), \underline{\xi})_{,i}| \times I))$ and $(|\Lambda^*(\lambda(\partial\sigma), \underline{\xi})_{,i}| \times I)$ are equal. d'_t extends to an isotopy d_t of $|(\xi, K_{,o}) \nmid (\lambda\partial\sigma)K| \times I$ rel $|(\xi, K_{,o}) \nmid (\lambda\partial\sigma)K| \times \{0, 1\} \cup$ $\cup |(\xi, K_{,o}) \nmid (\partial\sigma)K| \times I \cup |\lambda(\partial\sigma)| \times I$. Hence d_t is an allowable isotopy of the complementary flags $((\xi, K_{,o}) \nmid (\lambda\partial\sigma)K) \times I$ and $b(\partial'\mu \times I)$. This is the general step in our induction on $i = 1, \ldots, n$. At the end of the induction we have an isotopy e_t of $|(\xi, K_{,o}) \nmid (\lambda\partial\sigma)K| \times I$ rel $|(\xi, K_{,o}) \nmid (\lambda\partial\sigma)K| \times \{0, 1\} \cup |(\xi, K_{,o}) \nmid (\partial\sigma)K| \times I \cup |\lambda(\partial\sigma)| \times I$ such that the $(n, n-1)$-ns's $e_1 b(|\Lambda^*(\lambda(\partial\sigma), \underline{\xi})| \times I)$ and $|\Lambda^*(\lambda(\partial\sigma), \underline{\xi}| \times I$ are equal. Observe that $e_t b(\partial'\mu \times I)$ is complementary to the flag $((\xi, K_{,o}) \nmid (\lambda\partial\sigma)K) \times I$ for all t, and that $e_1 b(\partial'\mu \times I)$ is complementary to the stratification $\Lambda(\lambda(\partial\sigma), \underline{\xi}) \times I$. We can choose e_t to be level-preserving. Hence e_1 is the trace of an allowable isotopy e''_t of $|((\eta))(\partial\sigma)|$. Replacing b_t by $b''_t = e''_t \circ b_t$, we have isotoped $\partial'\mu$ to equal $\partial\mu$ through cS-block bundles complementary to the stratification $\Lambda(\lambda(\partial\sigma), \underline{\xi})$. b''_t extends to an isotopy $'b_t$ of the weak $(n+1, n)$-ns $|\Lambda(\lambda(\sigma), \underline{\xi})|$ rel $|\xi_{n+1, 0} \nmid \sigma K|$. So $\sigma\mu$, defined to be $'b_1('\mu)$ is still complementary to the stratification $\Lambda(\lambda(\sigma), \underline{\xi})$, and now extends $\rho\mu$ whenever $\rho \lneq \sigma$. This completes the inductive step in our construction of μ.

At the end of the induction we have a blocking μ/P of $((\eta))$ which is complementary to the flag $(\xi, K_{,o}) \nmid \lambda K$ and which is "locally" complementary to the stratification $\Lambda^*(\lambda, \underline{\xi})$, in the sense that each $\sigma\mu$ is complementary to $\Lambda^*(\lambda(\sigma), \underline{\xi})$. I leave the reader to check that μ is indeed complementary to $\Lambda^*(\lambda, \underline{\xi})$: consider, for example, the restriction $1'\mu/1'P$ of μ which covers $|\Lambda^*(\lambda, \underline{\xi})_{n+1,1}|$. $\underline{\xi}$ defines a flag structure $(''\xi, ''K_{,1})/''K = (\Lambda^*(\lambda, \underline{\xi})_{,1})$ on the family $(|\Lambda^*(\lambda, \underline{\xi})_{,1}|)$. Let $((''\eta))$ be a block decomposition blocked by $(''\xi, ''K_{,1})$, and let $''\mu/''P$ be a complementary blocking. We may assume that $((''\eta))$, and hence $''\mu$, respect each $|\Lambda^*(\lambda(\sigma), \underline{\xi})_{n+1,1}|$ as σ varies in L. By induction on a standard ordering of L, assume $''\mu$ and $1'\mu$ agree in $|\Lambda^*(\lambda(\partial\sigma), \underline{\xi})_{n+1,1}|$. By Lemma 8.2, there is an admissable isotopy of $|\Lambda^*(\lambda(\sigma), \underline{\xi})_{n+1,1}|$ rel $|\Lambda^*(\sigma, \underline{\xi})_{n+1,1}|$ which carries $''\mu$ isomorphically to $1'\mu$ in $|\Lambda^*(\lambda(\sigma), \underline{\xi})_{n+1,1}|$. It follows from the proof of that lemma that we may keep our isotopy fixed on $|\Lambda^*(\lambda(\partial\sigma), \underline{\xi})_{n+1,1}|$. By the standard procedure, this isotopy extends to an admissable isotopy of $|\Lambda^*(\lambda, \underline{\xi})_{n+1,1}|$ rel $|\Lambda^*(L, \underline{\xi})_{n+1,1}| = |1'P|$. So by induction there is an admissable isotopy of $|\Lambda^*(\lambda, \underline{\xi})_{n+1,1}|$ rel $|\Lambda^*(L, \underline{\xi})_{n+1,1}|$ which carries $1'\mu$ isomorphically to $''\mu$. By Lemma 8.1, $1'\mu$ is complementary to $(''\xi, ''K_{,1})$. Details are left to the reader, who will also see that the index 1 has played no part and is just a notational convenience for dealing with the general case, in which we consider the flag $(\Lambda^*(\lambda, \underline{\xi})_{,i})$. Hence μ is complementary to the stratification $\Lambda(\lambda, \underline{\xi})$.

Next we have to arrange that μ extends the given cS-block bundle ν^*. $(n+1, 0)\nu^*$ is complementary to $\Lambda((0,0)\nu^*, \underline{\xi})$; and by construction of μ there is a restriction μ^*/P^* of μ which is also complementary to this stratification. We need an admissable isotopy f'_t of $\Lambda((0,0)\nu^*, \underline{\xi})$ rel $|P^*|$ such that $f'_1 : \mu^* \approx (n+1, 0)\nu^*$ (after subdividing these cS-block bundles if necessary). We have already given the argument that shows how to do this: one starts with an admissable isotopy f''_t of the flag $(\xi^*, K^*_{,o}) \pitchfork \nu^* K^*$ rel $|P^*|$ which carries μ^* to $(n+1, 0)\nu^*$, and then adjusts f'_t till it respects the $(n, n-1)$-ns $\Lambda^*((0, 0)\nu^*, \underline{\xi})$; then the adjusted isotopy f'_t will do. Now f'_t extends to an isotopy of the weak $(n+1, n)$-ns $|\Lambda(\lambda, \underline{\xi})|$ rel $|P|$; then $f_t\mu$ is complementary to $\Lambda(\lambda, \underline{\xi})$ for all t. Set $\nu^{\sim}/Q^{\sim} = f_1(\mu/P)$.

Finally, set $M^{\wedge} = cl[M - |\xi_{n+1, 0}|]$, $Y^{\wedge} = Y \cap M^{\wedge}$, $X^{\wedge} = X \cap M^{\wedge}$, $X^{\wedge}* = (X^* \cap X^{\wedge}) \cup K'_{n, o}$. Then bdy $Y^{\wedge}$ has a normal cS-block bundle $\nu^{\wedge}*/Q^{\wedge}*$ in bdy M defined by $'\nu^{\sim}/'Q^{\sim} \cup \{(n+1, i)\nu^* : i = 1, \ldots, n\}$; and $\nu^{\wedge}*$ is complementary to $\sigma_o(\underline{\xi}^*) \cup \Lambda^*_o(\underline{\xi})$. By inductive hypothesis, Proposition 8.3(n-1), $\nu^{\wedge}*$ extends to a normal cS-block bundle $\nu^{\wedge}/Q^{\wedge}$ of $Y^{\wedge}$ in M complementary to $\sigma_o(\underline{\xi})$. Then $\nu = \nu^{\sim} \cup \nu^{\wedge}$ is complementary to $\underline{\xi}$. The inductive proof of Proposition 8.3(n-1) is complete.

<u>Corollary 1.</u> Given X, $Y \subset M$, (X), $\underline{\xi}/\underline{K}$ and ν^* as in the hypotheses of Proposition 8.3. Let ν/Q and ν'/Q' be normal cS-block bundles of Y in M which are both complementary to $\underline{\xi}$ and which both extend ν^*. Then there is an isotopy f_t of M rel bdy $M \cup Y$ such that $f_1 : \nu \approx \nu'$ (after subdividing these cS-block bundles if necessary) and $f_t\nu$ is complementary to $\underline{\xi}$ for all t.

We apply Proposition 8.3 to $X \times I$, $Y \times I \subseteq M \times I$, the stratification $\underline{\xi} \times I$ of $(X) \times I$ in M, and the normal cS-block bundle $\nu^{\sim}/Q^{\sim}$ of bdy $(Y \times I)$ in bdy$(M \times I)$ defined by $\nu \times 0 \cup \nu^* \times I \cup \nu' \times 1$. Then $\nu^{\sim}$

extends to a normal cS-block bundle $\hat{\nu}$ of $Y \times I$ in $M \times I$ complementary to $\underline{\xi} \times I$. The proof now continues as in the proof of Lemma 8.1: there is an isomorphism $\hat{\nu} \approx \nu \times I$ (after subdividing both) which can be used as the trace of the required isotopy. Details are left to the reader.

Corollary 2. Let $\underline{L}$ be an n-stratification, $\underline{K} < \underline{L}$ a restriction. Assume that $|K_{i,i}|$ is a proper submanifold of $|L_{i,i}|$ for $i = 0, \ldots, n$ (and not just an embedded manifold). Then $|K_{n,}|$ has a cS-block bundle neighbourhood ν/Q in $|L_{n,}|$ which is complementary to $\underline{L}$.

One can check that the proof of Proposition 8.3 applies. Or, let $f : |L_{n,}| \hookrightarrow M$ be an embedding into a large-dimensional manifold such that $f^{-1}\text{bdy } M = \text{bdy } |L_{n,}|$. Then $(|L_{n,}|, \ldots, |L_{o,}|)$ is a variety filtration of $|L_{n,}|$ in M, so there is a stratification $\underline{\eta}/\underline{L}$ of $(|L_{n,}|)$ in M. Let $\underline{\xi}/\underline{K} = \underline{\eta} \restriction \underline{K}$. Then $|\xi_{n+1,}| \perp \underline{\eta}$. The proof of Proposition 8.3 shows that $|\xi_{n+1,}|$ has a normal cS-block bundle neighbourhood ν'/Q' in M which is complementary to $\underline{\eta}$. Then $\nu/Q = \nu' \restriction Q' \langle |K_{n,}| \rangle$ is the required cS-block bundle.

Proposition 8.4(n). Given $X, Y \subseteq M$ such that M is a manifold, X a submanifold of M. Let $(Y_n, \ldots, Y_o)$ be a variety filtration of Y in M, and let ξ/K be a normal cS-block bundle of X in M such that $(Y) \pitchfork \xi$. Then there is a stratification $\underline{\eta}/\underline{L}$ of (Y) in M which is complementary to ξ; so in particular, $X \perp \underline{\eta}$. Further, given that $(Y^*) = (Y) \cap \text{bdy } M$ is $< \text{bdy}_M (Y)$, and given a stratification $\underline{\eta}^*/\underline{L}^*$ of (Y^*) in bdy M

which is complementary to ξ, then we may choose $\underline{\eta}$ to extend $\underline{\eta}^*$.

We use induction on n. Proposition 8.3(0) is just Rourke and Sanderson's [19, II, Proposition 4.10]. Now assume Proposition 8.4(n-1): $(Y) \cap X$ is a variety filtration of $Y \cap X$ in X. Let $\underline{N}$ be a normal (n+1, n)-ns for $(Y) \cap X$ in X; we may assume that $\underline{N}$ extends the normal (n+1, n)-ns $\{|\eta^*_{j,i}| \cap X\}$ of $(Y^*) \cap$ bdy X in bdy X. By subdividing ξ we may assume that every $N_{j,i}$ and $N^{\cdot}_{j,i}$ are covered by subcomplexes (j, i)K and (j, i)$^{\cdot}$K of K. Then $\{|\xi \restriction (j,i)K|\}$ defines a normal (n+1, n)-ns $\xi\underline{N}$ for $(Y) \cap |\xi|$ in $|\xi|$ which respects $\xi^{\cdot}$ and which extends the (n+1, n)-ns $|\underline{\eta}^*| \cap |\xi|$. Since $(Y) \pitchfork \xi$, $\xi\underline{N}$ extends to a normal (n+1, n)-ns $''\underline{N}$ of (Y) in M, which we may assume extends $|\underline{\eta}^*|$. $\underline{\eta}$ will be constructed as a blocking of $''\underline{N}$.

$\xi \restriction (n+1, 0)K$ satisfies comp 1 and 2, so there is a complementary flag $(\mu, \lambda_n, \ldots, \lambda_1)/P$. We may assume that (μ, λ) extends $(r^*, L^*_{,0}) \restriction \xi L^*_{0,0}$, where we write $\xi L^*_{0,0}$ for $L^*_{0,0} \langle \xi \restriction (\text{bdy } K \cap (0,0)K) \rangle$.
Let $\underline{\mu}'/\underline{P}'$ be an (n+1, n)-ns completion of (μ, λ) which extends $\Lambda(\xi L^*_{0,0}, \underline{\eta}^*)$. We have the normal (n, n-1)-ns's $|\Lambda^*(oP, \underline{\mu}')|$ and $\{N_{j,i} \cap N^{\cdot}_{n+1,0}\}$ of $(\lambda)^{\cdot} \restriction oP$ in $\mu^{\cdot} \restriction oP$, which agree in bdy X. So there is an isotopy a'_t of $(\mu, \lambda)^{\cdot} \restriction oP$ rel $|(n+1, 0)K \cap \text{bdy } K|$ such that $a'_1 : |\Lambda^*(oP, \underline{\mu}')| \approx \{N_{j,i} \cap N^{\cdot}_{n+1,0}\}$. a'_t extends to an isotopy a_t of $(M, \{(Y), X, |\xi|, \xi^{\cdot}\})$ rel bdy $M \cup Y_0$. Hence $a_1(\mu, \lambda)$ is still complementary to $\xi \restriction (n+1, 0)K$. Henceforth I shall suppress a_t and assume that $|\Lambda^*(oP, \underline{\mu}')|$ already is the ns $\{N_{j,i} \cap N^{\cdot}_{n+1,0}\}$.

Now $X \pitchfork \underline{\mu}'$. So by Proposition 8.3 (more precisely, by its proof) there is a cS-block bundle structure $\xi^{\sim}/K^{\sim}$ on $(|\xi|, X)$ which is complementary to $\underline{\mu}'$; and we may assume that $\xi^{\sim}$ extends $\xi \restriction \text{bdy } K$. We are only concerned with $\xi^{\sim} \restriction (n+1, 0)K^{\sim}$, which is a cS-block bundle complementary to $\underline{\mu}'$. Hence $\xi^{\sim} \restriction (n+1, 0)K^{\sim}$ and $\xi \restriction (n+1, 0)K$ are both complementary to the flag $(\mu, \lambda)/P$, and agree in bdy M. By Lemma 8.2 there is an admissable isotopy b_t' of (μ, λ) rel $|(n+1, 0)K^{\sim}| \cup$ $\cup\ |\xi \restriction ((n+1, 0)K \cap \text{bdy } K)|$ such that $b_1' : \xi^{\sim} \restriction (n+1,0)K^{\sim} \approx \xi \restriction (n+1,0)K$ (after subdividing $\xi^{\sim}$ and ξ if necessary). Then $b_1'\underline{\mu}'$ is complementary to $\xi \restriction (n+1, 0)K$; and it follows that the ns's $|\Lambda_o^*(b_1'\underline{\mu}')|$ and $\{\xi N_{j,i} \cap \xi N'_{n+1,0}\}$ are equal.

We have the weak $(n+1, n)$-ns $\underline{N}' = \{''N_{j,o} : j = 0, \dots, n+1\} \cup$ $\cup \{''N_{j,i} \cap {''N'_{n+1,0}} : i = 1, \dots, n,\ j = i, \dots, n+1\}$, which extends $|\underline{\mu}'|$ and $|\Lambda_o(\underline{\eta}^*)|$. Hence $\underline{\mu}'$ and $\Lambda_o(\underline{\eta}^*)$ extend to a blocking $\underline{\eta}'/\underline{L}'$ of $\underline{N}'$.

Finally set $M^{\wedge} = \text{cl}[M - {''X_{n+1,0}}]$, $X^{\wedge} = X \cap M^{\wedge}$, $(Y^{\wedge}) = (Y) \cap M^{\wedge}$, $\xi^{\wedge}/K^{\wedge} = \xi \restriction K\langle X^{\wedge}\rangle$. Then $\text{bdy}_M(Y^{\wedge}) = (Y^* \cap Y^{\wedge}) \cup L_n'^{\bullet}$ and has a stratification in bdy M $\underline{\eta}^{*\wedge} = \mathcal{C}_o(\underline{\eta}^*) \cup \Lambda_o^*(\underline{\eta}')$ which is complementary to $\xi^{\wedge}$. By inductive hypothesis, Proposition 8.4(n-1), $\underline{\eta}^{*\wedge}$ extends to a stratification $\underline{\eta}^{\wedge}/\underline{L}^{\wedge}$ of $(Y^{\wedge})$ in M which is complementary to $\xi^{\wedge}$. Then $\underline{\eta}/\underline{L} = \underline{\eta}' \cup \underline{\eta}^{\wedge}$ is the required stratification of (Y) in M. This completes the inductive proof of Proposition 8.4.

<u>Corollary 1.</u> Given $X, Y \subseteq M$, (Y), ξ and η^* as in the hypotheses of the proposition. Let $\underline{\eta}/\underline{L}$ and $\underline{\eta}'/\underline{L}'$ be stratifications of (Y) in M

which are both complementary to ξ and which both extend $\underline{\eta}^*$. Then there are subdivisions $'\underline{\eta}/'\underline{L}$ of $\underline{\eta}$, $'\underline{\eta}'/'\underline{L}'$ of $\underline{\eta}'$ and an isotopy f_t of $(M, \{(Y), X\})$ rel bdy M such that $f_1 : '\underline{\eta} \approx '\underline{\eta}'$ and $f_t\underline{\eta}$ is complementary to ξ for all t.

The proof is similar to the proof of the Corollary to Proposition 8.3, and is left to the reader. It should be noticed that one cannot in general keep (Y) fixed by f_t. This can be done, however, if $\underline{L} = \underline{L}'$.

Corollary 2. Let ξ/K be a cS-block bundle, and let $K_0 < \ldots < K_n = K$ be subcomplexes of K such that $(|K_n|, \ldots, |K_0|)$ is a variety filtration of $|K|$. Hence $(|\xi \upharpoonright K_n|, \ldots, |\xi \upharpoonright K_0|)$ is a variety filtration of $|\xi|$. Then there is an n-stratification $\underline{L}$ on $(|\xi|)$ which is complementary to ξ; so in particular, $|K|$ is covered by a restriction $J < L$.

One can check that the proof of Proposition 8.4 applies to this situation. Or, as in the proof of Corollary 2 to Proposition 8.3, let $f : |K| \hookrightarrow M$ be an embedding into a large-dimensional manifold such that $f^{-1}\text{bdy}\, M = \text{bdy}\,|K|$. Then ξ extends to a cS-block bundle ξ'/K' with $|K'| = M$, and $(|K_n|, \ldots, |K_0|)$ is a variety filtration of $|K|$ in $|K'|$. $|\xi'|$ is a manifold, and $|\xi| \perp \xi'$. So the proof of Proposition 8.4 shows there is an $(n+1, n)$-stratification $\underline{\eta}/\underline{L}$ of $(|\xi|)$ in $|\xi'|$ which is complementary to ξ'. Then $\underline{L}$ is the required n-stratification.

Putting together Propositions 8.3 and 8.4 gives:

Theorem 8.5. Given $X, Y \subseteq M$ with M a manifold and X a submanifold of M; then X is block transverse to Y if and only if Y is block transverse to X.

From Corollary 2 to Proposition 8.3 and Corollary 2 to Proposition 8.4 we infer:

Theorem 8.6. A p.ℓ. map $f : X \to Y$ is stratifiable if and only if there are a large-dimensional disk D and a factorization $f = p \circ g$ such that $p : Y \times D \to Y$ is the projection, $g : X \to Y \times D$ is an embedding, and gX has a cS-block bundle neighbourhood in $Y \times D$.

It is not hard to see (using the method of proof of Lemma 5.2) that a p.ℓ. embedding $f : X \hookrightarrow Y$ is stratifiable if and only if the pair (Y, X) is equivalent to a pair $(|\xi|, |K|)$ for some cS-block bundle ξ/K - in other words, one can cancel the disk D in the statement of the theorem.

Appendix

In order to develop the theory of flags and of decompositions, we introduce the notion of an "Ω cone-block bundle", where Ω is a partially ordered set. Our treatment of Ω cbb's now follows Rourke and Sanderson's [19, I and II]. Thus our Result X, Y. Z is a reformulation of [19, X, Result Y. Z]. It is usually left to the reader to reformulate Rourke and Sanderson's proof; we confine ourselves to details that did not arise in the case of block bundles.

Let Ω be a finite partially-ordered set with a unique minimal element 0 and a unique maximal element α. An Ω cone-block bundle $\{\xi\}$ over a cell complex K is a family $\{\xi_\omega : \omega \in \Omega\}$ of cone-block bundles over K such that:

Ω cbb 1 if $\omega < \omega'$ in Ω, then $|\xi_\omega| \subseteq |\xi_{\omega'}|$;

Ω cbb 2 $\xi_0 = K$;

Ω cbb 3 for each $\sigma \in K$, there are a compact polyhedron with a family of subpolyhedra indexed by Ω, say $(F_\alpha(\sigma), \{F_\omega(\sigma)\})$, and a block structure $h : \xi_\alpha(\sigma) \to \sigma \times cF_\alpha(\sigma)$ which, for every $\omega \in \Omega$, restricts to a block structure $h : \xi_\omega(\sigma) \to \sigma \times cF_\omega(\sigma)$.

We write: $\{\xi\}/K$ is an Ω cbb.

We leave the reader to define the restriction $\{\xi\}|L$ of an Ω cbb $\{\xi\}/K$ to a subcomplex $L < K$.

If $\{\xi\}/K$ and $\{\eta\}/L$ are Ω cbb's, then an Ω cbb isomorphism

$f : \{\xi\} \approx \{\eta\}$ is a p.ℓ. isomorphism $f : |\xi_\alpha| \to |\eta_\alpha|$ which restricts to a cone-block bundle isomorphism $f : \xi_\omega \approx \eta_\omega$ for every $\omega \in \Omega$.

Given Ω as before; then a family of compact polyhedra $(F_\alpha, \{F_\omega\})$ indexed by Ω is an Ω family if $F_0 = \emptyset$; $F_\omega \subseteq F_{\omega'}$ whenever $\omega < \omega'$; and $F_\omega \cap F_{\omega'} =$ some $F_{\omega''}$ whenever $\omega, \omega' \in \Omega$. An Ω cbb $\{\xi\}/K$ is a c$\{F\}$-block bundle if for every $\sigma \in K$, $(F_\alpha(\sigma), \{F_\omega(\sigma)\}) = (F_\alpha, \{F_\omega\})$. The proof of Lemma 1.7 shows that given an Ω cbb $\{\xi\}/K$ and $\sigma \in K$, then $(F_\alpha(\sigma), \{F_\omega(\sigma)\})$ is unique. By Theorem I, 1.1 (below), any $\{\xi\}$ is a disjoint union of some c$\{F^i\}$-block bundles $\{\xi^i\}/K^i$, where the K^i are the components of K. If $\{F\}$ is an Ω family and K is a cell complex, then we have the trivial c$\{F\}$-bb $K \times c\{F\}$ (see Chapter 3 for the product of blocks). A trivialization of an Ω cbb $\{\xi\}/K$ is an isomorphism with some trivial c$\{F\}$-bb over K; $\{F\}$ is then uniquely determined.

Let $\{\xi\}/K$ be an Ω cbb, σ a cell of K. Then a chart for $\{\xi\}$ at σ is a trivialization $h(\sigma)$ of $\{\xi\} \vdash \langle\sigma\rangle$. An atlas for $\{\xi\}$ is a family $\{h(\sigma) : \sigma \in K, h(\sigma)$ is a chart at $\sigma\}$.

Let $(F_\alpha, \{F\})$ be an Ω-family. Then we have the disjunction $\{\mathfrak{F}_\omega : \omega \in \Omega\}$ of F_α defined by $\mathfrak{F}_\omega = F_\omega - \cup\{F_{\omega'} : F_{\omega'} \subsetneq F_\omega\}$. Henceforth we shall always assume that $\{\mathfrak{F}_\omega\}$ is a variety of F_α. Thus if $\{\xi\}/K$ is an Ω cbb, and we write $\{\xi\}$ as a disjoint union of c$\{F^i\}$-block bundles, then the disjunction $\{\mathfrak{F}^i_\omega\}$ associated to each $\{F^i\}$ is a variety.

Theorem I, 1.1(n). Let $\{\xi\}/K$ be an Ω cbb with $|K| \underset{p.\ell.}{\approx} D^n$ (an n-disk). Then $\{\xi\}$ is trivial.

We use induction on n. For $n = 0$, K is a single vertex, and $\{\xi\}$ is trivial by Ω cbb 3. Now assume Theorem I, 1.1(n-1). We mark with an asterisk the results that depend on this assumption.

Lemma I, 1.2. Given an Ω family $\{F\}$, cells σ^1 and σ^2, and a p.ℓ. isomorphism

$$f : (\partial\sigma^1 \times c\{F\}) \cup (\sigma^1 \times \{F\}) \cup \sigma^1 \longrightarrow$$

$$\longrightarrow (\partial\sigma^2 \times c\{F\}) \cup (\sigma^2 \times \{F\}) \cup \sigma^2$$

with $f^{-1}(\sigma^2) = \sigma^1$. Then f extends to a p.$\ell$. isomorphism $f' : \sigma^1 \times c\{F\} \to \sigma^2 \times c\{F\}$.

Pick points $x^j \in \text{int}\,\sigma^j$ for $j = 1, 2$, and regard $\sigma^j \times c\{F\}$ as the cone from x^j on $\partial\sigma^j * \{F\}$. Then the restriction of f to $g : \partial\sigma^1 * \{F\} \to \partial\sigma^2 * \{F\}$ extends to an isomorphism $g'' : \sigma^1 \times c\{F\} \to \sigma^2 \times c\{F\}$ by extending conically to x^1 and x^2. Now g'' need not equal f on σ^1; however $g''(\sigma^1) = \sigma^2$ and g'' equals f on $\partial\sigma^1$. So there is an isotopy h_t of σ^2 rel $\partial\sigma^2$ such that $h_1 \circ g''$ equals f on σ^1. The proof of Akin's Covering Isotopy Theorem [1] shows that h_t extends to an isotopy h''_t of $\sigma^2 \times c\{F\}$ rel $\partial\sigma^2 * \{F\}$. Then $f' = h''_1 \circ g''$ is the required isomorphism.

Proposition I, 1.3*. Let $\{\xi\}/K$ be an Ω cbb such that $|K| \underset{p.\ell.}{\approx} D^n$ and K has just one n-cell, σ^n. Let ρ^{n-1} be any (n-1)-cell in K, and let $L = K - \{\sigma, \rho\}$. Given a trivialization $t : \{\xi\}|L \approx L \times c\{F\}$. Then t extends to a trivialization $t'' : \{\xi\} \approx K \times c\{F\}$.

By Ω cbb 3 there is a block structure $h : \xi_\alpha(\sigma) \to \sigma \times cF_\alpha$

which restricts to a structure $h : \xi_\omega(\sigma) \to \sigma \times cF_\omega$ for all $\omega \in \Omega$. Let $V_\alpha = h^{-1}(\partial\sigma \times cF_\alpha \cup \sigma \times F_\alpha)$, $V_\omega = h^{-1}(\partial\sigma \times cF_\omega \cup \sigma \times F_\omega)$. Then $|\{\xi\} \restriction L|$ is a regular neighbourhood of $|L|$ rel bdy $|L|$ in $(V_\alpha, \{V_\omega\})$. Pick points $x \in \text{int}\ |L|$ and $y \in \text{int}\ \rho$. Then there is a p.ℓ. isomorphism $p : (V_\alpha, \{V_\omega\}) \to (x \cup y) * (\text{bdy}\ |L| * (F_\alpha, \{F_\omega\}))$ and we may assume p corresponds $|\{\xi\} \restriction L| \longleftrightarrow x * (\text{bdy}|L| * F_\alpha)$. Let $(W_\alpha, \{W_\omega\}) = p^{-1}(y * (\text{bdy}|L| * (F_\alpha, \{F_\omega\})))$; then $W_\alpha \cap |K| = \rho$. By Lemma I, 1.2 the p.ℓ. isomorphism $t : p^{-1}(\text{bdy}|L| * (F_\alpha, \{F_\omega\})) \longrightarrow$ $\longrightarrow \partial\rho * (F_\alpha, \{F_\omega\})$ together with the identity on ρ extend to an isomorphism $t'' : (W_\alpha, \{W_\omega\}) \to \rho \times c(F_\alpha, \{F_\omega\})$. This, together with t, gives an isomorphism (different from p)

$$t_1 : (V_\alpha, \{V_\omega\}) \longrightarrow (\partial\sigma \times c(F_\alpha, \{F_\omega\})) \cup (\sigma \times (F_\alpha, \{F_\omega\})).$$

The proof now follows that given by Rourke and Sanderson, except that one uses Cohen's Regular Neighbourhood Theorem (our Theorem 1.1) instead of the regular neighbourhood theorem they quote.

<u>Proposition I, 1.4*</u>. Any Ω cbb $\{\xi\}/K$ with $\dim|K| \leq n$ has an atlas.

Let $\{\xi\}/K$, $\{\eta\}/L$ be Ω cbb's. Then $\{\eta\}$ is a <u>subdivision</u> of ξ if η_ω/L is a subdivision of ξ_ω/K (as cone block bundles) for all $\omega \in \Omega$.

<u>Theorem I, 1.5*</u>. 1. Let $\{\xi\}/K$ be an Ω cbb with $\dim|K| \leq n$, and let L be a cell complex subdivision of K. Then there is a subdivision $\{\eta\}/L$ of $\{\xi\}$.

2. Let $\{\xi\}/K$ be an Ω cbb with $\dim|K| \leq n$. Given $K' < K$ and a

subdivision $\{\eta'\}/L'$ of $\{\xi\}|K'$. Set $\{\eta\}/L = (\{\xi\} - \{\xi\}|K) \cup \{\eta'\}$ (as a set of blocks. Then $\{\eta\}$ is an Ω cbb and subdivides $\{\xi\}$.

Theorem I, 16.*. Let $\{\xi\}/K$ and $\{\xi'\}/K'$ be Ω cbb's with $\dim|K| \leq n$. Assume $|K| \searrow |L|$ (collapses geometrically), where L is a subcomplex of K. Given a cell complex isomorphism $h : K \approx K'$ and an extension of h to an isomorphism of Ω cbb's $h' : \{\xi\}|L \approx \{\xi'\}|hL'$. Then h and h' extend to an Ω cbb isomorphism $h'' : \{\xi\} \approx \{\xi'\}$.

Corollary I, 1.7*. Given an Ω cbb $\{\xi\}/K$ such that $\dim|K| \leq n$ and $|K| \searrow 0$. Then $\{\xi\}$ is trivial.

The proofs are the same as in [8, I, §1.] In particular, Theorem I, 1.1(n) holds, and the inductive step is proved. Theorem I, 1.1 and the starred results now hold without restriction on n.

Corollary I, 1.8. Given Ω cbb's $\{\xi\}/K \times I$ and $\{\eta\}/K \times I$, a subcomplex $L < K$, and an isomorphism $h : \{\xi\}|(K\times 0 \cup L\times I) \approx \{\eta\}|(K\times 0 \cup L\times I)$ which is the identity on $K \times 0 \cup L \times I$. Then h extends to an isomorphism $h' : \{\xi\} \approx \{\eta\}$ which is the identity on $K \times I$.

Theorem I, 1.9. Let $\{\eta\}/L$ and $\{\eta'\}/L$ be subdivisions of an Ω cbb $\{\xi\}/K$. Then $\{\eta\} \approx \{\eta'\}$.

Given an Ω cbb $\{\xi'\}/K'$ and a cell complex K of which K' is a subdivision. For each $\sigma \in K$, let $h : \{\xi'\}|K'\langle\sigma\rangle \to K'\langle\sigma\rangle \times c\{F\}$ be a trivialization, given by Theorem I, 1.1. For each $\omega \in \Omega$ define a block $\xi_\omega(\sigma)$ so that $h : \xi_\omega(\sigma) \to \sigma \times cF_\omega$ is a block structure. Then

$\{\xi\} = \{\xi_\omega(\sigma) : \sigma \in K, \omega \in \Omega\}$ is an Ω cbb over K, called the amalgamation of $\{\xi'\}$ over K. Note that $\{\xi'\}$ is a subdivision of $\{\xi\}$.

Given Ω cbb's $\{\xi\}/K$ and $\{\eta\}/L$ with $|K| = |L|$. Then $\{\xi\}$ and $\{\eta\}$ are equivalent if there are subdivisions $\{\xi'\}/K'$ of $\{\xi\}$, $\{\eta'\}/L'$ of $\{\eta\}$, and an isomorphism $h : \{\xi'\} \approx \{\eta'\}$ which is the identity on $|K|$. We write: $\{\xi\} \sim \{\eta\}$.

Now for each cell complex K and each fixed Ω family $\{F\}$, let $I(K, \{F\})$ be the set of isomorphism classes of $c\{F\}$-bb's $\{\xi\}/K$ (here all isomorphisms are understood to be the identity on K). For each polyhedron X and each fixed Ω family $\{F\}$, let $I(X, \{F\})$ be the set of equivalence classes of $c\{F\}$-bb's $\{\xi\}/K$ with $|K| = X$.

Theorem I, 1.10. The map $a : I(K, \{F\}) \to I(|K|, \{F\})$, which assigns to each isomorphism class of Ω cbb's its equivalence class, is a bijection.

Let u/X be an element of $I(X, \{F\})$, and let $Y \subset X$ be a subpolyhedron. Take a cell complex triangulation K of X in which Y is covered by a subcomplex $L < K$. Represent u by $\{\xi\}/K \in I(K, \{F\})$. Then the equivalence class v/Y of $\{\xi\}|L$ depends only on u and Y, and is called the restriction of u to Y. We write: $v = u|Y$, and sometimes, if u is represented by $\{\xi'\}/K'$, we refer to v loosely as: $\{\xi'\}|Y$.

If Ω and Ω' are partially-ordered sets with unique minimal elements 0 and 0', and unique maximal elements α and α', then $\Omega \times \Omega'$ has the partial ordering: $\omega \times \varphi' < \chi \times \psi'$ if $\omega < \chi$ and $\varphi' < \psi'$. $0 \times 0'$ is the unique minimal element, and $\alpha \times \alpha'$ the unique maximal element of $\Omega \times \Omega'$. Now let $\{\xi\}/K$ be an Ω cbb, $\{\eta'\}/L'$ an Ω'cbb. Then $\{\xi \times \eta'\}$, defined as $\{\beta \times \gamma' : \beta \in \{\xi\}, \gamma \in \{\eta'\}\}$ (see Chapter 3 for the definition of the block $\beta \times \gamma'$), is an $\Omega \times \Omega'$ cbb over $K \times L'$. $\{\xi \times \eta'\}$ is the cartesian <u>product</u> of $\{\xi\}$ and $\{\eta'\}$. This product gives rise to a well-defined product on equivalence classes, $I(X, \{F\}) \times I(Y, \{F'\}) \to I(X \times Y, \{F * F'\})$. Note that if $\{F\}$ is an Ω-family and $\{F'\}$ an Ω'-family such that the associated disjunctions $\{\mathfrak{F}\}$ and $\{\mathfrak{F}'\}$ are varieties, then the disjunction associated to the $\Omega \times \Omega'$ family $\{F * F'\}$ is again a variety.

Hence we can define the Whitney <u>sum</u> of equivalence classes u/X and v/X as $u \times v \pitchfork \Delta$, where Δ, the diagonal of $X \times X$, is identified with X by $(x, x) \longleftrightarrow x$. We write: $u \oplus v/X$.

Given $f : X \to Y$ a p.ℓ. map, and u/Y. Then the <u>induced</u> class f^*u/X is defined as $X \times u \pitchfork \Gamma f$, where X/X is regarded as the only element of $I(X, \{\emptyset\})$, and Γf, the graph of f, is identified with X by $(x, f(x)) \longleftrightarrow x$.

<u>Theorem I, 1.11.</u> Given $u/(X \times I)$; then $u = (u \pitchfork (X \times 0)) \times I$.

<u>Corollary I, 1.12.</u> Given homotopic maps $f, g : X \to Y$, and a class u/Y Then $f^*u = g^*u$.

Let Ω be a partially ordered set with unique minimal element 0 and unique maximal element α. Define $\Omega+$ to have as underlying set the disjoint union of Ω and an element λ, with partial ordering generated by the partial ordering on Ω together with: $\alpha < \lambda$. If $(F_\alpha, \{F_\omega\})$ is an Ω family and S^q is a sphere of sufficiently large dimension, then there is a p.ℓ. embedding $i : F_\alpha \hookrightarrow S^q$, and the $\Omega+$ family $(S^q, \{iF_\alpha, iF_\omega\})$ is unique up to p.ℓ. isomorphism. We abbreviate it to: $(S^q, \{F_\alpha, F_\omega\})$.

<u>Proposition I, 2.1.</u> Given an Ω cbb $\{\xi\}/K$ with fibre $(F_\alpha, \{F_\omega\})$. Then there is an $\Omega+$ cbb $\{\xi^*\}/K$ with fibre $(S^q, \{F_\alpha, F_\omega\})$, for some sufficiently large q, such that ξ^*_λ/K is the trivial cS^q-block bundle over K, and $\{\xi^*_\alpha, \xi^*_\omega\}/K = \{\xi\}/K$.

Take $q \geq 2(\dim|\xi_\alpha| + 1)$. Then we have $q \geq 2(\dim F_\alpha + 1)$, so there is a unique embedding of F_α in S^q. The proof of [8, I, Proposition 2.1] gives a p.ℓ. embedding $h : |\xi_\alpha| \hookrightarrow |K| \times D^{q+1}$ such that: $h : |K| \approx |K| \times 0$ is the obvious map, where 0 is an interior point of D^{q+1}; $h^{-1}(\sigma \times D^{q+1}) = |\xi_\alpha(\sigma)|$ for every $\sigma \in K$. Let $(V, \{V_\omega(\sigma)\})$ be a regular neighbourhood of $|K| \times 0$ in $(|K| \times D^{q+1}, \{h \mid \xi_\omega(\sigma)|\})$, where ω ranges over Ω, and σ over K. Then we can define an $\Omega +$ cbb $\{''\xi^*\}/K$ by the rules:

$|''\xi^*_\lambda(\sigma)| = (\sigma \times D^{q+1}) \cap V;$

$\mathcal{G}''\xi^*_\lambda(\sigma) = \sigma;$

$\partial''\xi^*_\lambda(\sigma) = (\partial\sigma \times D^{q+1}) \cap V;$

$''\xi^{*\,\bullet}_\lambda(\sigma) = (\sigma \times D^{q+1}) \cap \mathrm{fr}\, V;$

$|''\xi^*_\omega(\sigma)| = V_\omega(\sigma)$;

for all $\sigma \in K$ and for all $\omega \in \Omega$. Let $\{''\xi\}/K$ be the Ω cbb $\cup\{''\xi^*_\omega : \omega \in \Omega\}$. Then there is an isomorphism of Ω cbb's $f : \{''\xi\} \approx \{\xi\}$ which is the identity on K. (This follows from the fact that $(|''\xi_\alpha|, \{|''\xi_\omega(\sigma)|\})$ is a regular neighbourhood of $|K|$ in $(|\xi_\alpha|, \{|\xi_\omega(\sigma)|\})$, which is itself a weak regular neighbourhood of $|K|$.) Under this identification f, $\{''\xi^*\}$ becomes the required $\Omega+$ cbb $\{\xi^*\}/K$.

Now let $f : L \to K$ be a cell complex map. Then we have $f \times \text{id.} : |L| \times D^{q+1} \to |K| \times D^{q+1} = |\xi^*_\lambda|$. Define the Ω cbb $\{f^\# \xi\}/L$ to have blocks:

$|f^\# \xi_\omega(\tau)| = (f\times\text{id.})^{-1}|\xi_\omega| \cap (\tau \times D^{q+1})$;

$\mathcal{B}(f^\# \xi_\omega(\tau)) = \tau$;

$\partial(f^\# \xi_\omega(\tau)) = (f\times\text{id.})^{-1}|\xi_\omega| \cap (\partial\tau \times D^{q+1})$;

$f^\# \xi^\bullet_\omega(\tau) = (f\times\text{id.})^{-1}(\xi^\bullet_\omega) \cap (\tau \times D^{q+1})$;

for all $\tau \in L$, and for all $\omega \in \Omega$. Then $\{f^\# \xi\}$ has fibre $\{F\}$ (we are assuming $\{\xi\}$ is a $c\{F\}$-bb.)

<u>Proposition I, 1.13.</u> If $\{\xi\}/K$ represents $u \in I(X, \{F\})$, then $\{f^\# \xi\}/L$ represents f^*u.

<u>Corollary I, 1.14.</u> $(f\circ g)^* = g^*\circ f^*$, for any maps f and g.

<u>Theorem I, 4.1.</u> Given a $c\{F\}$-bb $\{\xi\}/K$, and subdivisions $\{\xi'\}/K'$, $\{\eta'\}/K'$ of $\{\xi\}$. Let $L < K$ have induced subdivision $L' < K'$ and be

such that $\{\xi'\}|L' = \{\eta'\}|L'$. Then there is an isotopy f_t of $|\xi_\alpha|$ rel $|K|$ such that:

$f_1 : \{\xi'\} \approx \{\eta'\}$;

$f_t(\{\xi'\})$ is a subdivision of $\{\xi\}$ for all t;

f_t is the identity on $|\xi_\alpha \nmid L|$.

Further, if $\Omega^* \subseteq \Omega$ is a subset, and $\xi'_{\omega^*} = \eta'_{\omega^*}$ for every $\omega^* \in \Omega^*$, then we may add: f_t is the identity on every $|\xi'_{\omega^*}|$. (Note that $\xi'_\omega = \eta'_\omega$ whenever $\omega <$ some $\omega^* \in \Omega^*$.)

The proof of this extra condition, and others like it, is an application of Akin's Covering Isotopy Theorem [1], quoted as our Theorem 1.6.

Theorem I, 4.4. 1. Let $(X_n, \ldots, X_o)$ be a stratified polyhedron, and let $(\xi)/K$, $(\eta)/K$ be n-flag neighbourhoods of X_o in $(X_n, \ldots, X_1)$. Then there is an isotopy f_t of $(X_n, \ldots, X_o)$ rel X_o such that

$f_1 : (\xi) \approx (\eta)$.

2. Further if $X^* <$ bdy X, and (ξ) and (η) respect X^*, with $(\xi)|K^* = (\eta)|K^*$, then we may add:

f_t is the identity on X^*.

3. Further if for some $m \leq n$, $(\xi_m, \ldots, \xi_1)/K = (\eta_m, \ldots, \eta_1)/K$ then we may add: f_t is the identity on X_m.

We shall also need a slightly different version of [8, I, Theorem 4.4]:

Theorem I, 4.A. 1. Let $\{\xi\}/K$ and $\{\eta\}/K$ be c{F} -bb's such that $|K|$ is a manifold, $|\xi_\omega| = |\eta_\omega|$ and $\dot{\xi}_\omega = \dot{\eta}_\omega$ for all $\omega \in \Omega$. Then there is an isotopy f_t of $(|\xi_\alpha|, \{|\xi_\omega|, \dot{\xi}_\alpha\})$ rel $|K|$ such that $f_1 : \{\xi\} \approx \{\eta\}$.

2. Further, if $K^* < \text{bdy}\ K$ and $\{\xi\}|K^* = \{\eta\}|K^*$, then we may add: f_t is the identity on $|\xi_\alpha|K^*|$.

3. Further, if $\Omega^* \subset \Omega$ is a subset, and $\xi_{\omega^*} = \eta_{\omega^*}$ for every $\omega^* \in \Omega^*$, then we may add: f_t is the identity on $\cup\{|\xi_{\omega^*}|\}$. (Note that $\xi_\omega = \eta_\omega$ whenever $\omega <$ some $\omega^* \in \Omega^*$.)

Proposition II, 4.4. Any block decomposition over a disk is trivial. Hence a block decomposition is in face a decomposition.

Theorem II, 4.5. Given block decompositions $((\eta_1)) = [(\xi_1); (\zeta_1)]/K$ and $((\eta_2)) = [(\xi_2); (\zeta_2)]/K$, and isomorphisms $g : \xi_1 \approx \xi_2$, $h : \zeta_1 \approx \zeta_2$. Then g and h extend to an isomorphism $f : ((\eta_1)) \approx ((\eta_2))$.

Proposition II, 4.6. Given flags $(\xi)/K$ and $(\zeta)/K$. Then we have block decompositions $[(\xi) \times K; K \times (\zeta)]/K \times K$ and $[K \times (\zeta); (\xi) \times K]/K \times K$.

Let P be a cell complex triangulation of $|(\xi)|$ in which every block of (ξ) and its rim are covered by subcomplexes. Let Q be a similar triangulation of $|(\zeta)|$. Then $P \times (\zeta)/P \times K$ and $(\xi) \times Q/K \times Q$ are complementary blockings of $[(\xi) \times K;\ K \times (\zeta)]$ and of $[K \times (\zeta); (\xi) \times K]$.

Theorem II, 4.7. Given flags $(\xi)/K$, $(\zeta)/K$. Then there is a block decomposition $(\xi) \oplus (\zeta) = [(\xi_1); (\zeta_1)]$ such that $(\xi_1) \approx (\xi)$ and $(\zeta_1) \approx (\zeta)$; and $(\xi) \oplus (\zeta) = [(\zeta_1); (\xi_1)]$ is also a block decomposition.

For if $t : (\xi) \times (\zeta) \approx (\zeta) \times (\xi)$ is the isomorphism (of Ω cbb's, where $\Omega = \{0, \ldots, p\} \times \{0, \ldots, q\}$) which exchanges the factors, then t induces an equivalence $t^* : (\xi) \oplus (\zeta) \sim (\zeta) \oplus (\xi)$, by restriction to the diagonal Δ of $|K \times K|$. Then in the notation of the previous proposition, $(\xi) \oplus (\zeta)$ has the complementary blockings $P \times (\zeta) \upharpoonright \Delta$ and $(\xi) \times Q \upharpoonright \Delta$.

<u>Theorem II, 4.8.</u> If $((\eta)) = [(\xi); (\zeta)]/K$ is a block decomposition, then there is an isomorphism $h : ((\eta)) \approx (\xi) \oplus (\zeta)/K$; and $((\eta))$ has the block decomposition structure $[(\zeta); (\xi)]$ given by $h^{-1} \circ t^* \circ h$.

If $(\mu)/R$ is a blocking of $[(\xi); (\zeta)]$, then we can choose h to give an isomorphism $h : (\mu) \approx P \times (\zeta) \upharpoonright \Delta$. Then $(h^{-1} \circ t^* \circ h)^{-1}((\xi) \times Q \upharpoonright \Delta)$ is a blocking of $[(\zeta); (\xi)]$ complementary to (μ).

Bibliography and Related Reading

[1] E. Akin - Manifold phenomena in the theory of polyhedra, Trans. Amer. Math. Soc. 143 (1969), 413-473.

[2] E. Akin - Cone complexes and transverse cellular maps, to appear.

[3] M. A. Armstrong - Transversality for polyhedra, Ann. of Math. 86 (1967), 172-191.

[4] M. A. Armstrong and E. C. Zeeman - Piecewise linear transversality, Bull. Amer. Math. Soc. 73 (1967), 184-188.

[5] A. J. Casson - Fibrations over spheres, Topology 6 (1967), 489-499.

[6] M. M. Cohen - Simplicial structures and transverse cellularity, Ann. of Math. 85 (1967), 218-245.

[7] M. M. Cohen - A general theory of relative regular neighborhoods, Trans. Amer. Math. Soc. 136 (1969), 189-229.

[8] A. Haefliger - Knotted spheres and related geometric topics, report to the I.C.M., Moscow, 1966.

[9] A. Haefliger and C. T. C. Wall - Piecewise linear bundles in the stable range, Topology 4 (1965), 209-214.

[10] H. Hironaka - Resolution of singularities of an algebraic variety over a field of characteristic zero: I, II, Ann. of Math. 79 (1964), 109-326.

[11] M. A. Kervaire and J. Milnor - Groups of homotopy spheres I, Ann. of Math. 77 (1963), 504-537.

[12] W. B. R. Lickorish - The piecewise linear unknotting of cones, Topology 4 (1965), 67-91.

[13] W. B. R. Lickorish and L. C. Siebenmann - Regular neighbourhoods and the stable range, Trans. Amer. Math. Soc. 139 (1969), 207-230.

[14] S. Łojasiewicz - Triangulation of semi-analytic sets, Pisa, Scuola Norm. Sup. Ann. Sci. Fis. e Mat., 18 (1964), 449-474.

[15] C. McCrory - Ph.D. thesis, Brandeis University, 1971.

[16] J. Milnor - Singular points of complex hypersurfaces, Ann. of Math. Studies, No. 61, Princeton, 1968.

[17] H. R. Morton - Joins of polyhedra, Topology, 9 (1970), 243-250.

[18] C. P. Rourke - The Hauptvermutung according to Sullivan I and II, lecture notes, Institute for Advanced Study, 1968.

[19] C. P. Rourke and B. J. Sanderson - Block bundles I, II and III, Ann. of Math. 87 (1968), 1-28, 256-278 and 431-483.

[20] C. P. Rourke and B. J. Sanderson - An embedding without a normal microbundle, Invent. Math. 3 (1967), 293-299.

[21] C. P. Rourke and B. J. Sanderson - Decompositions and the relative tubular neighbourhood conjecture, Topology 9 (1970), 225-229.

[22] C. P. Rourke and B. J. Sanderson - Δ-sets I and II, Quart. Journ. Math., to appear.

[23] N. E. Steenrod and D. B. A. Epstein - Cohomology operations, Ann. of Math. Studies, No. 50, Princeton, 1962.

[24] D. A. Stone - A counterexample in block bundle theory, Topology 9 (1970), 11-12.

[25] D. P. Sullivan - Triangulating homotopy equivalences, Ph. D. thesis, Princeton, 1965.

[26] D. P. Sullivan - Geometry topology, part I: Localization, Periodicity and Galois symmetry, lecture notes, M.I.T., 1970.

[27] R. Thom - Quelques propriétés globales des variétés differentiables, Comment. Math. Helvet. 28 (1954), 17-56.

[28] R. Thom - Ensembles et morphismes stratifiés, Bull. Amer. Math. Soc. 75 (1969), 240-284.

[29] H. Whitney - Local properties of analytic varieties, in: Differential and combinatorial topology, a symposium in honor of Marston Morse (ed. Cairns), 205-244, Princeton, 1965.

[30] R. E. Williamson - Cobordism and combinatorial manifolds, Ann. of Math. 83 (1966), 1-33.

[31] E. C. Zeeman - Dihomology III, Proc. Cam. Phil. Soc. (3) 13 (1963), 155-183.

[32] E. C. Zeeman - Seminar on combinatorial topology, mimeographed notes, I.H.E.S., Paris, 1963.

Vol. 146: A. B. Altman and S. Kleiman, Introduction to Grothendieck Duality Theory. II, 192 pages. 1970. DM 18,–

Vol. 147: D. E. Dobbs, Cech Cohomological Dimensions for Commutative Rings. VI, 176 pages. 1970. DM 16,–

Vol. 148: R. Azencott, Espaces de Poisson des Groupes Localement Compacts. IX, 141 pages. 1970. DM 16,–

Vol. 149: R. G. Swan and E. G. Evans, K-Theory of Finite Groups and Orders. IV, 237 pages. 1970. DM 20,–

Vol. 150: Heyer, Dualität lokalkompakter Gruppen. XIII, 372 Seiten. 1970. DM 20,–

Vol. 151: M. Demazure et A. Grothendieck, Schémas en Groupes I. (SGA 3). XV, 562 pages. 1970. DM 24,–

Vol. 152: M. Demazure et A. Grothendieck, Schémas en Groupes II. (SGA 3). IX, 654 pages. 1970. DM 24,–

Vol. 153: M. Demazure et A. Grothendieck, Schémas en Groupes III. (SGA 3). VIII, 529 pages. 1970. DM 24,–

Vol. 154: A. Lascoux et M. Berger, Variétés Kähleriennes Compactes. VII, 83 pages. 1970. DM 16,–

Vol. 155: Several Complex Variables I, Maryland 1970. Edited by J. Horváth. IV, 214 pages. 1970. DM 18,–

Vol. 156: R. Hartshorne, Ample Subvarieties of Algebraic Varieties. XIV, 256 pages. 1970. DM 20,–

Vol. 157: T. tom Dieck, K. H. Kamps und D. Puppe, Homotopietheorie. VI, 265 Seiten. 1970. DM 20,–

Vol. 158: T. G. Ostrom, Finite Translation Planes. IV. 112 pages. 1970. DM 16,–

Vol. 159: R. Ansorge und R. Hass. Konvergenz von Differenzenverfahren für lineare und nichtlineare Anfangswertaufgaben. VIII, 145 Seiten. 1970. DM 16,–

Vol. 160: L. Sucheston, Constributions to Ergodic Theory and Probability. VII, 277 pages. 1970. DM 20,–

Vol. 161: J. Stasheff, H-Spaces from a Homotopy Point of View. VI, 95 pages. 1970. DM 16,–

Vol. 162: Harish-Chandra and van Dijk, Harmonic Analysis on Reductive p-adic Groups. IV, 125 pages. 1970. DM 16,–

Vol. 163: P. Deligne, Equations Différentielles à Points Singuliers Reguliers. III, 133 pages. 1970. DM 16,–

Vol. 164: J. P. Ferrier, Seminaire sur les Algebres Completes. II, 69 pages. 1970. DM 16,–

Vol. 165: J. M. Cohen, Stable Homotopy. V, 194 pages. 1970. DM 16,–

Vol. 166: A. J. Silberger, PGL_2 over the p-adics: its Representations, Spherical Functions, and Fourier Analysis. VII, 202 pages. 1970. DM 18,–

Vol. 167: Lavrentiev, Romanov and Vasiliev, Multidimensional Inverse Problems for Differential Equations. V, 59 pages. 1970. DM 16,–

Vol. 168: F. P. Peterson, The Steenrod Algebra and its Applications. A conference to Celebrate N. E. Steenrod's Sixtieth Birthday. VII, 317 pages. 1970. DM 22,–

Vol. 169: M. Raynaud, Anneaux Locaux Henséliens. V, 129 pages. 1970. DM 16,–

Vol. 170: Lectures in Modern Analysis and Applications III. Edited by C. T. Taam. VI, 213 pages. 1970. DM 18,–

Vol. 171: Set-Valued Mappings, Selections and Topological Properties of 2^X. Edited by W. M. Fleischman. X, 110 pages. 1970. DM 16,–

Vol. 172: Y.-T. Siu and G. Trautmann, Gap-Sheaves and Extension of Coherent Analytic Subsheaves. V, 172 pages. 1971. DM 16,–

Vol. 173: J. N. Mordeson and B. Vinograde, Structure of Arbitrary Purely Inseparable Extension Fields. IV, 138 pages. 1970. DM 16,–

Vol. 174: B. Iversen, Linear Determinants with Applications to the Picard Scheme of a Family of Algebraic Curves. VI, 69 pages. 1970. DM 16,–

Vol. 175: M. Brelot, On Topologies and Boundaries in Potential Theory. VI, 176 pages. 1971. DM 18,–

Vol. 176: H. Popp, Fundamentalgruppen algebraischer Mannigfaltigkeiten. IV, 154 Seiten. 1970. DM 16,–

Vol. 177: J. Lambek, Torsion Theories, Additive Semantics and Rings of Quotients. VI, 94 pages. 1971. DM 16,–

Vol. 178: Th. Bröcker und T. tom Dieck, Kobordismentheorie. XVI, 191 Seiten. 1970. DM 18,–

Vol. 179: Seminaire Bourbaki – vol. 1968/69. Exposés 347-363. IV. 295 pages. 1971. DM 22,–

Vol. 180: Séminaire Bourbaki – vol. 1969/70. Exposés 364-381. IV, 310 pages. 1971. DM 22,–

Vol. 181: F. DeMeyer and E. Ingraham, Separable Algebras over Commutative Rings. V, 157 pages. 1971. DM 16,–

Vol. 182: L. D. Baumert. Cyclic Difference Sets. VI, 166 pages. 1971. DM 16,–

Vol. 183: Analytic Theory of Differential Equations. Edited by P. F. Hsieh and A. W. J. Stoddart. VI, 225 pages. 1971. DM 20,–

Vol. 184: Symposium on Several Complex Variables, Park City, Utah, 1970. Edited by R. M. Brooks. V, 234 pages. 1971. DM 20,–

Vol. 185: Several Complex Variables II, Maryland 1970. Edited by J. Horváth. III, 287 pages. 1971. DM 24,–

Vol. 186: Recent Trends in Graph Theory. Edited by M. Capobianco/J. B. Frechen/M. Krolik. VI, 219 pages. 1971. DM 18,–

Vol. 187: H. S. Shapiro, Topics in Approximation Theory. VIII, 275 pages. 1971. DM 22,–

Vol. 188: Symposium on Semantics of Algorithmic Languages. Edited by E. Engeler. VI, 372 pages. 1971. DM 26,–

Vol. 189: A. Weil, Dirichlet Series and Automorphic Forms. V, 164 pages. 1971. DM 16,–

Vol. 190: Martingales. A Report on a Meeting at Oberwolfach, May 17-23, 1970. Edited by H. Dinges. V, 75 pages. 1971. DM 16,–

Vol. 191: Séminaire de Probabilites V. Edited by P. A. Meyer. IV, 372 pages. 1971. DM 26,–

Vol. 192: Proceedings of Liverpool Singularities – Symposium I. Edited by C. T. C. Wall. V, 319 pages. 1971. DM 24,–

Vol. 193: Symposium on the Theory of Numerical Analysis. Edited by J. Ll. Morris. VI, 152 pages. 1971. DM 16,–

Vol. 194: M. Berger, P. Gauduchon et E. Mazet. Le Spectre d'une Variété Riemannienne. VII, 251 pages. 1971. DM 22,–

Vol. 195: Reports of the Midwest Category Seminar V. Edited by J.W. Gray and S. Mac Lane. III, 255 pages. 1971. DM 22,–

Vol. 196: H-spaces – Neuchâtel (Suisse) – Août 1970. Edited by F. Sigrist. V, 156 pages. 1971. DM 16,–

Vol. 197: Manifolds – Amsterdam 1970. Edited by N. H. Kuiper. V, 231 pages. 1971. DM 20,–

Vol. 198: M. Hervé, Analytic and Plurisubharmonic Functions in Finite and Infinite Dimensional Spaces. VI, 90 pages. 1971. DM 16,–

Vol. 199: Ch. J. Mozzochi, On the Pointwise Convergence of Fourier Series. VII, 87 pages. 1971. DM 16,–

Vol. 200: U. Neri, Singular Integrals. VII, 272 pages. 1971. DM 22,–

Vol. 201: J. H. van Lint, Coding Theory. VII, 136 pages. 1971. DM 16,–

Vol. 202: J. Benedetto, Harmonic Analysis on Totally Disconnected Sets. VIII, 261 pages. 1971. DM 22,–

Vol. 203: D. Knutson, Algebraic Spaces. VI, 261 pages. 1971. DM 22,–

Vol. 204: A. Zygmund, Intégrales Singulières. IV, 53 pages. 1971. DM 16,–

Vol. 205: Séminaire Pierre Lelong (Analyse) Année 1970. VI, 243 pages. 1971. DM 20,–

Vol. 206: Symposium on Differential Equations and Dynamical Systems. Edited by D. Chillingworth. XI, 173 pages. 1971. DM 16,–

Vol. 207: L. Bernstein, The Jacobi-Perron Algorithm – Its Theory and Application. IV, 161 pages. 1971. DM 16,–

Vol. 208: A. Grothendieck and J. P. Murre, The Tame Fundamental Group of a Formal Neighbourhood of a Divisor with Normal Crossings on a Scheme. VIII, 133 pages. 1971. DM 16,–

Vol. 209: Proceedings of Liverpool Singularities – Symposium II. Edited by C. T. C. Wall. V, 280 pages. 1971. DM 22,–

Vol. 210: M. Eichler, Projective Varieties and Modular Forms. III, 118 pages. 1971. DM 16,–

Vol. 211: Théorie des Matroïdes. Edité par C. P. Bruter. III, 108 pages. 1971. DM 16,–

Vol. 212: B. Scarpellini, Proof Theory and Intuitionistic Systems. VII, 291 pages. 1971. DM 24,–

Vol. 213: H. Hogbe-Nlend, Théorie des Bornologies et Applications. V, 168 pages. 1971. DM 18,–

Vol. 214: M. Smorodinsky, Ergodic Theory, Entropy. V, 64 pages. 1971. DM 16,–